Make:

JUMPSTARTING
the Arduino 101

INTERACTING WITH A COMPUTER
THAT LEARNS

Yining Shi | Sagar Mohite

MAKER MEDIA
SAN FRANCISCO, CA

Published by Maker Media, Inc., 1700 Montgomery Street, Suite 240, San Francisco, CA 94111

Maker Media books may be purchased for educational, business, or sales promotional use. Online editions are also available for most titles (*safaribooksonline.com*). For more information, contact our corporate/institutional sales department: 800-998-9938 or *corporate@oreilly.com*.

Publisher: Roger Stewart
Editor: Patrick DiJusto
Copy Editor: Elizabeth Welch, Happenstance Type-O-Rama
Proofreader: Scout Festa, Happenstance Type-O-Rama
Interior Designer and Compositor: Maureen Forys, Happenstance Type-O-Rama
Cover Designer: Maureen Forys, Happenstance Type-O-Rama
Indexer: Valerie Perry, Happenstance Type-O-Rama

All the circuit and component diagrams in this book are created using Fritzing (*http://fritzing.org/home*).

August 2017: First Edition

Revision History for the First Edition
2017-08-28 First Release

See *oreilly.com/catalog/errata.csp?isbn=9781680454550* for release details.

Safari® Books Online

Safari Books Online is an on-demand digital library that delivers expert content in both book and video form from the world's leading authors in technology and business. Technology professionals, software developers, web designers, and business and creative professionals use Safari Books Online as their primary resource for research, problem solving, learning, and certification training. Safari Books Online offers a range of plans and pricing for enterprise, government, education, and individuals. Members have access to thousands of books, training videos, and prepublication manuscripts in one fully searchable database from publishers like O'Reilly Media, Prentice Hall Professional, Addison-Wesley Professional, Microsoft Press, Sams, Que, Peachpit Press, Focal Press, Cisco Press, John Wiley & Sons, Syngress, Morgan Kaufmann, IBM Redbooks, Packt, Adobe Press, FT Press, Apress, Manning, New Riders, McGraw-Hill, Jones & Bartlett, Course Technology, and hundreds more. For more information about Safari Books Online, please visit us online.

How to Contact Us

Please address comments and questions to the publisher:

Maker Media
1700 Montgomery St.
Suite 240
San Francisco, CA 94111

You can send comments and questions to us by email at
books@makermedia.com.

Maker Media unites, inspires, informs, and entertains a growing community of resourceful people who undertake amazing projects in their backyards, basements, and garages. Maker Media celebrates your right to tweak, hack, and bend any Technology to your will. The Maker Media audience continues to be a growing culture and community that believes in bettering ourselves, our environment, our educational system—our entire world. This is much more than an audience, it's a worldwide movement that Maker Media is leading. We call it the Maker Movement.

To learn more about Make: visit us at makezine.com. You can learn more about the company at the following websites:

Maker Media: makermedia.com

Maker Faire: makerfaire.com

Maker Shed: makershed.com

DEDICATION

To my parents, for their endless support
—Yining Shi

CONTENTS

ACKNOWLEDGMENTS

I'd like to thank my co-author, Sagar Mohite, for spending numerous hours writing the book with me and for believing in me all the time; Tom Igoe, for his guidance, super-helpful suggestions, and feedback; editor Patrick DiJusto, for helping me and encouraging me throughout the whole process; and copy editor Elizabeth Welch and designer Maureen Forys for doing such an amazing job.

—*Yining Shi*

It's been a real pleasure working on this book, and I want to thank Yining Shi for inviting me to collaborate. I also want to thank my parents for their support and everyone who helped make this book a reality.

—*Sagar Mohite*

PREFACE

Arduino has taken over the maker movement over the last decade as one of the most accessible hardware and software platforms for creatives and makers. Arduino 101 is one of the newest boards from Arduino. It combines the ease-of-use of boards like Arduino Uno with powerful features, such as Bluetooth, motion sensing, and gesture recognition.

Over the last two decades, the web has also evolved to be the most accessible and engaging form of media. In this book you'll also explore the possibilities of creating web-based interfaces to communicate with Arduino 101 via the recently drafted Web Bluetooth API. You'll also learn how to use the accelerometer and gyroscope built into the 101, along with Intel Curie's pattern matching engine, to train the board and build your own media player.

This hands-on introduction to Arduino 101, complete with source code and walkthroughs, will help you start prototyping your projects right away.

USING THE CODE EXAMPLES

The source code for all experiments in this book is included inline with explanations where necessary. Additionally, it is hosted on my GitHub account *https://github.com/yining1023/Jumpstarting-the-Arduino-101)* and is distributed under the MIT License.

1

What Is Arduino?

Arduino is an open source platform that enables creatives, artists, makers, and engineers to prototype interactive physical computing projects. The Arduino ecosystem can be broadly divided into two parts: Arduino boards and the Arduino software, or the integrated development environment (IDE).

Arduino closely incorporates the open source philosophy, which, broadly speaking, is all about making the source and the design documents of products free to use, share, and distribute under open licenses. The idea is to make knowledge more accessible and therefore encourage collaboration and exchange of ideas among people. Arduino has released all the original design files (EAGLE CAD) for the Arduino hardware under a Creative Commons Attribution Share-Alike license, which allows for both personal and commercial derivative works, as long as they credit Arduino and release their designs under the same license. The source code for the software is released under various GNU licenses.

HARDWARE

Many of you might already be familiar with an Arduino board. An Arduino board is a small computer that can be connected to various sensors like thermometers, accelerometers, and photoresistors to measure the world around us. It can also be programmed to control LEDs, as well as actuators like motors, to affect the world around us.

The Arduino hardware ecosystem offers a wide variety of products that you can use for prototyping. Every Arduino product is designed to cater to a specific type of application. For example, though the Arduino Uno is a great board for entry-level projects, you will probably benefit from the form factor and the size of the Arduino Gemma board for wearable projects. The Arduino Yun, on the other hand, is great for projects that require wireless communication and are in the *Internet of Things* (IoT) domain.

In addition to boards, Arduino offers modules, shields, and kits. Modules are smaller versions of some of the popular boards; shields are accessories that you can connect to your boards to bring in more features; and kits are combination packs of components, boards, and accessories created for specific purposes.

All the experiments in this book are based on the Arduino 101, which is a cool new entry-level board with a lot of features packed in. I recommend referring to some of the other books in the Make series to learn more about Arduino and to get started with other boards.

SOFTWARE

Now that you have an idea of what hardware Arduino has to offer, let's look at all the things that are possible with these boards. This is where the Arduino software comes in. The Arduino software complements the hardware by providing an easy-to-use coding environment that you can take advantage of to program the board. A lot of people generally use what is colloquially called the *Arduino language* to write programs (also known as *sketches*) for their boards. But the Arduino language is simply a set of C/C++ functions that are written in a way that makes it very easy to learn. However, you are free to code in C/C++, since your Arduino sketches are ultimately compiled with a C/C++ compiler (avr-g++).

You can read more on the build process for your Arduino sketches here: *https://github.com/arduino/Arduino/wiki/Build-Process*.

ARDUINO 101

Arduino 101 is one of the newer boards from Arduino. Although the 101 offers features similar to other boards like the Uno, what really sets it apart are the various onboard modules. Together they provide a variety of benefits, like efficiency, performance, wireless connectivity, and motion sensing, in the same compact form factor as the Uno.

FIGURE 1.1: Arduino 101, front side

FIGURE 1.2: Arduino 101, back side

This new board houses an Intel Curie module, which offers better performance at a lower power footprint. The module has two 32-bit microcontroller units (MCUs)—an x86 Intel Quark processor and an ARC EM4 processor—along with 384 kB of flash memory and 80 kB of static random access memory (SRAM). These onboard MCUs combine a variety of new technologies, including wireless communication via Bluetooth Low Energy, a six-axis motion sensor with an accelerometer, and a gyroscope.

> NOTE Bluetooth Low Energy is also referred to as BLE, Bluetooth Smart, Bluetooth 4.0, or Bluetooth LE throughout this book.

The 101 can, therefore, be easily paired with other Bluetooth devices, and can detect various motions, such as shocks, taps, free falls, and even gestures. In fact, the Quark MCU on the

Curie module features a neural network that opens up many machine learning possibilities.

If you are curious about how the Arduino core framework interfaces with the real-time operating system on the Curie module, I encourage you to refer to the documentation and tools open sourced by Curie's Open Developer Kit (*https://software .intel.com/en-us/node/674972*).

Let's explore two of the main features of 101: BLE and motion sensing with the inertial measurement unit.

Bluetooth Low Energy

One of the things I really like about BLE is the accessibility and ubiquity of this technology. The fact that BLE modules ship with most smartphones these days makes it a very appealing technology when it comes to making things that can talk with phones and other Bluetooth-enabled devices. Unlike Bluetooth Classic, BLE is extremely suitable for low-power applications where only small amounts of data need to be exchanged. This makes it perfect for IoT projects, where a device has to run on battery power for extended periods of time. Chapter 3, "Exploring Bluetooth LE on the Arduino 101," talks more in depth about how BLE works.

Motion Sensing with the Inertial Measurement Unit

One of the many reasons why I like the Arduino 101 is that it has a lot of features offered right out of the box at a very affordable price. As I mentioned earlier, the Curie module features a six-axis motion sensor, also known as an inertial measurement unit (IMU), which consists of a three-axis accelerometer and a three-axis gyroscope module. This is great because a three-axis accelerometer and three-axis gyroscope module by itself might cost you about one-third of the price of the 101 if you buy it separately.

IMUs are often used to detect various gestures and patterns, as you will read about in Chapter 4, "Exploring Motion Sensors on the Arduino 101," and Chapter 5, "Exploring Pattern Matching and Machine Learning on Intel Curie."

Specifications

Table 1.1 shows some of the technical specifications for the 101, as stated by Arduino.

TABLE 1.1: Technical specifications

MICROCONTROLLER	INTEL CURIE
Operating voltage	3.3V (5V tolerant I/O)
Input voltage (recommended)	7–12V
Input voltage (limit)	7–17V
Digital I/O pins	14 (of which 4 provide pulse width modulation [PWM] output)
PWM digital I/O pins	4
Analog input pins	6
DC current per I/O pin	20 mA (milliamps)
Flash memory	196 kB
SRAM	24 kB
Clock speed	32 MHz
LED_BUILTIN	13
Features	Bluetooth LE, six-axis accelerometer/gyro
Length	68.6 mm (millimeters)
Width	53.4 mm
Weight	34 grams

Input and Output

Let's get familiar with the some of the important peripherals of the 101 board.

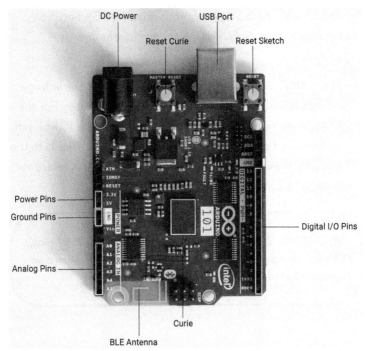

DC Power   USB Port

Reset Curie   Reset Sketch

Power Pins

Ground Pins

Digital I/O Pins

Analog Pins

Curie

BLE Antenna

FIGURE 1.3: The main components and inputs, outputs on Arduino 101

Power

There are various ways to power your 101:

* The DC power jack input can be in the 7–12V range, and it accepts a 2.1 mm center positive plug. Most AC-to-DC wall adapters already have this. You can also use a battery with such a plug.

* The Vin pin also accepts a 7–12V power supply. If you are using a battery, the leads can be inserted into the GND (ground) pin and the VIN pin of the power connector.

* The USB port accepts a regulated 5V input with a USB connector.

* The 5V and the 3.3V pins output regulated 5V and 3.3V voltage, respectively. The maximum current draw is 1500 mA.

* The IOREF pin on all Arduino boards provides the voltage reference with which the microcontroller operates. For the 101, you will read a value of 3.3V on this pin. The IOREF pin is often used to connect external modules and shields to the Arduino; these modules and shields then use this reference voltage to configure their power source.

> **WARNING** Do not supply voltage to the 5V or 3.3V pins, because doing so bypasses the internal regulator and can damage your board if it is not sufficiently regulated.

Digital I/O Pins

The 101 has 14 general-purpose digital input/output (I/O) pins. These pins can be configured to be either input pins or output pins. The `pinMode()` function is used for configuring whether you want the pin to behave as an input pin or as an output pin. By default, all digital pins are set up as INPUT.

Input pins are considered to be in a *high-impedance* state. Such a configuration enables the circuit to be sensitive to very small changes in the magnitude of the current flowing through the input pin. So you can use a pin in an INPUT mode to detect changes in the current flow from an external circuit change.

Pins configured as OUTPUT are considered to be in a *low-impedance* state. This sets up the circuit to behave in the exact opposite way as the input state. An output pin can provide a significant amount of current to other circuits connected to it. You can then write up code that determines when to provide a current to an external circuit.

Alternatively, you can also configure a pin to operate in an INPUT_PULLUP mode. In this configuration, the board uses an internal pull-up resistor between the pin and GND. While in an INPUT mode, if there is nothing connected to the pin, it will arbitrarily read either HIGH or LOW. Such a pin is called a *floating pin*. If you connect this pin to +5V, there are chances of shorting. In such a case you can use the internal pullup resistor of the board by setting the pin mode to INPUT_PULLUP using code.

> **WARNING** In most cases, it's a good idea to connect output pins to other devices with 70 Ω or 1 kΩ (1000 Ω) resistors. This can prevent short circuits and damage to your board.

Pulse Width Modulation

Four of the digital pins—3, 5, 6, and 9—can also be used for pulse width modulation (PWM). PWM is a method used to simulate analog signals with a digital medium. It can allow you to control the voltage provided to inertial loads like motors or capacitors.

With digital signals, a signal can either have a high state (1) or a low state (0). Alternating between a low and a high state at a fixed period of time can be used to generate a square wave, as shown in Figure 1.4. The ratio of the signal's on state to period of time is called the *duty cycle*. So a 30 percent duty cycle means the signal is on 30 percent of the time and off 70 percent of the time.

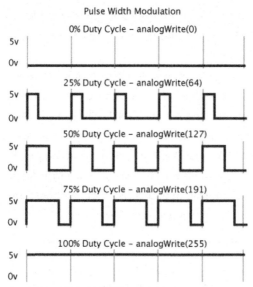

FIGURE 1.4: Pulse width modulation

You can use PWM to modulate the average voltage supplied to a device. For example, an LED can be dimmed by approximately 50 percent using PWM without reducing the supply voltage. To do this, you can set up the PWM signal with a 50 percent duty cycle. This means that the voltage provided to the LED will be HIGH for half the time and LOW for the other half. The voltage will oscillate between these two states at the frequency of the PWM square wave, which for Arduino is 500 Hz. Because this frequency is higher than what your eyes can perceive, you will see a decrease in brightness of the LED.

With pulse width modulation, you can simulate an analog output on any of the PWM pins with the analogWrite() method.

Analog Input Pins

The 101 has six analog input pins. Analog signals provided to these pins are converted to digital signals using an internal analog-to-digital converter (ADC) circuit. The `analogRead()` function can be used to read an analog signal from these pins. We usually use such pins to read analog outputs from sensors like a photoresistor or a force sensor.

Additionally, the analog pins have all the functionalities of the digital pins, so you can use them just like digital I/O pins in case you need more of these ports.

2

Getting Started

The Arduino IDE allows you to write programs and upload them to your Arduino board. Two editors are available: the online Arduino Web Editor and the Arduino Desktop IDE. If you prefer to use the Arduino Web Editor, you can follow the instructions found at *https://create.arduino.cc/projecthub/Arduino_Genuino/ getting-started-with-the-arduino-web-editor-4b3e4a*. In this chapter, we are going to use the Arduino Desktop IDE.

INSTALLING THE ARDUINO DESKTOP IDE

Go to the download page for the Arduino website (*https://www .arduino.cc/en/Main/Software*), choose the correct version of the Arduino IDE for your computer, and download it.

Download the Arduino IDE

FIGURE 2.1: Download the Arduino IDE page

Installing on macOS/OS X

Download the IDE for your Mac. Once you download the zip file, extract the contents and open the extracted directory. Copy the app package named `Arduino` and paste it into your `Applications` directory. Then you can launch the Arduino application.

> NOTE Copying the app into your `Applications` directory is an optional step, and you can choose not to do it.

Installing on Windows

You can install the IDE on your Windows machine in several ways. The easiest way is to install it directly from the EXE file available to download and then follow the instructions (Figure 2.2). The

EXE file also contains all the drivers you will need to connect your board with your Windows machine. If you get a warning during the process, allow the driver installation for a smooth install.

FIGURE 2.2: Installing Arduino IDE on Windows

If you don't have sufficient privileges to install the Arduino IDE with the executable file, you can download the zip file for Windows and extract and run the bundled application. If you install in this manner, you'll have to install all the correct drivers manually.

Installing on Linux

Here are the steps for installing on Linux (Figure 2.3):

1. Download and save the latest version of the IDE for your 32-bit, 64-bit, or ARM-based Linux OS.

2. Extract the compressed package into a suitable directory.

3. Open a Terminal window and cd into the extracted directory (which will probably be named arduino-1.x.x unless you specified a different name while extracting the files).

4. Run the installation script by typing ./install.sh in your Terminal and pressing Enter.

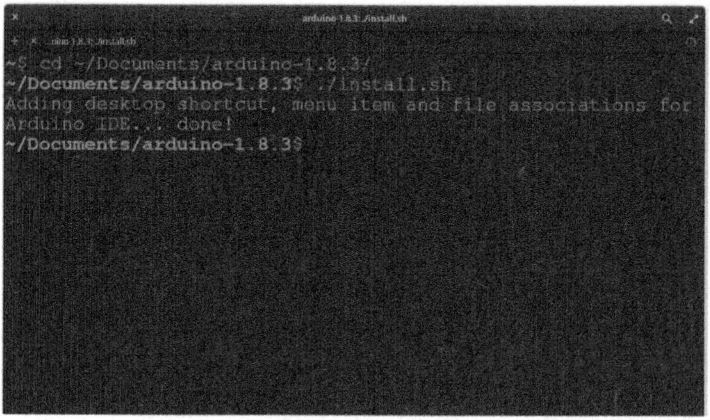

FIGURE 2.3: Installing Arduino IDE on Linux

Now you can just run the installed Arduino IDE app from your
Applications folder.

Connecting Your Arduino 101

This part of the process is somewhat different from using other
Arduino boards. Connect your 101 to your computer with a USB
cable. If this is the first time you're connecting your 101, the Arduino IDE should guide you to installing the necessary libraries for
the board. If not, then open your Arduino IDE and click Tools >
Boards > Boards Manager (Figure 2.4). Search for "Intel Curie
Boards by Intel," select it, and install the latest version (Figure 2.5).

FIGURE 2.4: Opening Boards Manager in the Arduino Desktop IDE

FIGURE 2.5: Installing the Intel Curie Boards in Boards Manager

After you install the Intel Curie Boards, you can select the Arduino 101 board by going to Tools > Boards > Arduino/Genuino 101 (Figure 2.6). Then you must select the correct port to communicate with Arduino 101. Click Tools > Port and select a port that reflects the USB port you have used to plug in your 101 (Figure 2.7). If you are on a Mac, you can see options like /dev/cu.usbmodem, /dev/tty.usbmodem, or both. You can select either of them. They both refer to a USB port of your Mac.

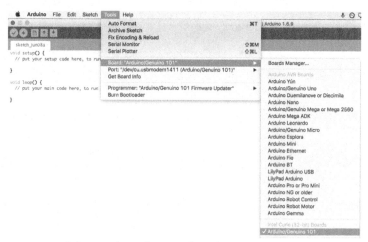

FIGURE 2.6: Selecting the Arduino 101 board

FIGURE 2.7: Choose the right port.

Now that you have successfully set up your 101 with the Arduino IDE, it's time to start programming the board to do some fun things.

UNDERSTANDING AN ARDUINO SKETCH

Writing programs for an Arduino board is easy. An Arduino program is a unit of code that is written to govern the behavior of your board. It is commonly referred to as a *sketch*. Let's start by exploring the two basic functions that constitute an Arduino sketch: the `setup()` function and the `loop()` function:

THE setup() FUNCTION

The setup() function is called once when the sketch starts. You can use it for different setup tasks like importing libraries, initializing global variables, and assigning pin modes.

THE loop() FUNCTION

The loop() function is called in a loop over and over again throughout the life cycle of the sketch. This is where most of the read and write operations happen, and most of the logic of your sketch will likely reside in this function.

Once you are done writing an Arduino sketch, you can compile and upload the sketch to your board.

The Verify button compiles your sketch to machine readable code (binaries).

During this time the compiler can detect any errors in code and warn you about them.

The Upload button compiles your sketch and uploads the compiled binaries to your board.

The Serial Monitor button opens up a separate Terminal window. The Serial Monitor is used to log incoming or outgoing communications over the serial port. You can also use it to print any messages while debugging your sketch.

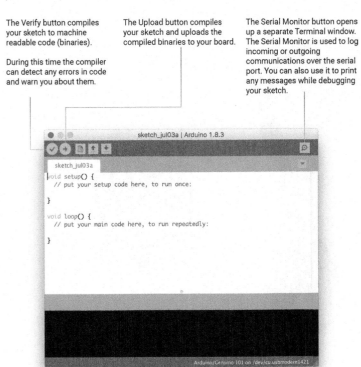

FIGURE 2.8: The Arduino IDE interface

We'll talk in detail about other functions available to you in the following chapters.

BLINKING AN LED

Blinking an LED is a great first project for setting up and testing your Arduino 101 board. In this experiment, you're going to connect a light-emitting diode (LED) to your Arduino 101, and program the board to turn the LED on and off every one second.

You'll need the following (Figure 2.9):

* Arduino 101 (with USB cables)

* An LED

FIGURE 2.9: What you'll need for the Blinking an LED project

Circuit Assembly

Arduino 101 comes with an LED on the board, which is marked as L. It is connected to pin number 13. You can use the `digitalWrite()` function to write a HIGH or a LOW state to pin 13. With the HIGH setting, the board will provide +3.3V to this pin and light up the internal LED. Similarly, a LOW state will ground this pin (0V) and turn off the LED.

If you want to use an external LED, you can do so by connecting its anode (positive, longer leg) to one of the digital pins, and its cathode (negative, shorter leg) to the GND pin. The connection is shown in Figure 2.10.

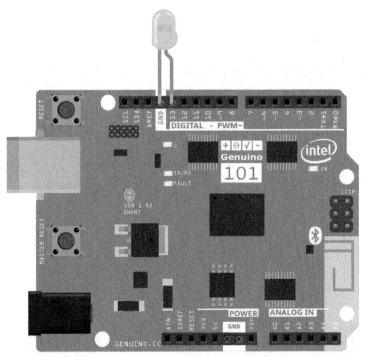

FIGURE 2.10: Circuit and component diagram of the Blinking an LED project

CODE

Now that you have the circuit wired up, you can write a program
that will make the Arduino turn the LED on and off every second.
The Arduino IDE has a built-in Blink example. Open the IDE and
choose File > Examples > 01.Basics > Blink (Figure 2.11). This will
open the source code for the Blink sketch (Figure 2.12).

FIGURE 2.11: Opening the Blink example

FIGURE 2.12: Blink example

The complete code can also be found here: *https://github.com/yining1023/Jumpstarting-the-Arduino-101*.

This is a good time to walk through the different parts of the code to understand it. Let's start with the `setup()` function. In your `setup()` function, you are doing one thing: configuring pin 13 as an OUTPUT pin. You can use the `pinMode()` function to set this up. The `pinMode(pin, mode)` function accepts two arguments: the pin number and the mode. As discussed previously, your `pinMode` can be INPUT, OUTPUT, or INPUT_PULLUP. The pin number in this case is an integer representing the pin you have connected with your LED. You are saving this value in a variable called `ledPin` to reuse that value in other parts of your sketch code.

```
int ledPin = 13;

void setup()
{
  pinMode(ledPin, OUTPUT);
}
```

The next part of the code is the `loop()` function, which will be called over and over throughout the life of the sketch. Most of our logic for this example will reside in this function. You'll begin by writing a HIGH state to your pin.

The `digitalWrite(pin, value)` function takes two arguments: a pin number and a value. The value can be either HIGH or LOW. The constants HIGH and LOW are both defined globally for your convenience, and you can use them directly. Let's set the value to HIGH and turn on our LED.

```
void loop()
{
  digitalWrite(ledPin, HIGH);
}
```

Next we call the `delay(milliseconds)` function. This takes one argument: the number of milliseconds that you want to pause your sketch.

```
void loop()
{
  digitalWrite(ledPin, HIGH);
  delay(1000);
}
```

You then set the pin to a LOW state again and wait for another second.

```
void loop()
{
  digitalWrite(ledPin, HIGH);   // sets the LED on
  delay(1000);                  // waits for a second
  digitalWrite(ledPin, LOW);    // sets the LED to off
  delay(1000);                  // waits for a second
}
```

Since the `loop()` function will repeat itself indefinitely, the whole cycle will go on and your LED will start blinking. Here is the complete source code for our Blinking example:

```
int ledPin = 13;              // LED connected to digital pin 13

void setup()                  // setup() function is called when a
                              // sketch starts
{
  pinMode(ledPin, OUTPUT);    // sets the digital pin as output
}

void loop()                   // loop() function will be called
                              // over and over again
{
  digitalWrite(ledPin, HIGH); // sets the LED on
  delay(1000);                // waits for a second
  digitalWrite(ledPin, LOW);  // sets the LED off
  delay(1000);                // waits for a second
}
```

I encourage you to visit the Arduino Language Reference page (*https://www.arduino.cc/en/Reference/HomePage*) for in-depth explanations about the various functions, variables, and constants available.

Now, it's time to verify (or compile) your sketch and upload it to the board. Make sure you choose the right port and the right board, as we discussed earlier in the "Installing the Arduino Desktop IDE" section.

Click the Verify button in the upper-left corner of the IDE. If you see a Done Compiling message below the code window, that means the Arduino Desktop IDE converted your human-readable program to something that the Arduino board can understand— also called a *binary*.

At this point, click the Upload button next to the Verify button. If you see a Done Uploading message this time, the LED on your Arduino 101 board should start blinking. As you might notice, the built-in LED and the external LED will both be turned on/off at the same time, because they are both connected to pin 13.

Congratulations! You just finished your first project with Arduino 101. We can now dive into other features that make Arduino 101 really fun.

FINAL RESULT

Here are images showing our circuit with the LED blinking.

FIGURE 2.13: "Blinking an LED" final result 1 (LED is off)

FIGURE 2.14: "Blinking an LED" final result 2 (LED is on)

TROUBLESHOOTING

If you get any error message while verifying the program, read the error message and double-check your program again. If you couldn't upload the program successfully, make sure you chose the right board in Tools > Boards > Arduino/Genuino 101, and that you chose the right port (which differs depending on your computer's operating system). Check the "Installing the Arduino Desktop IDE" section for more information.

3

Exploring Bluetooth LE on the Arduino 101

Bluetooth Low Energy (BLE), originally started by Nokia as an in-house project called "Wibree," has grown to become a standard in wireless communication protocols. Operating on the same frequency as classic Bluetooth, BLE offers decreased power consumption while maintaining a similar range.

Whereas classic Bluetooth is great for streaming audio and other applications that have a high data throughput, Bluetooth LE is primarily meant for applications that need to exchange only small amounts of data at a slower rate. This makes BLE an ideal choice for communications among low-power connected devices like fitness trackers and smart watches.

CENTRAL VS. PERIPHERAL DEVICES

The network topology, and the way communication happens over Bluetooth LE, is one of the main reasons for its low power consumption. BLE's communication style is similar to a typical client-server model. You can set up a BLE-capable device as either a peripheral device (the equivalent of a server) or a central device (similar to a client). Think of the peripheral device as a server containing information. Information on a peripheral device is organized as a collection of services, each having a different set of characteristics, such as "heart rate" for a fitness service, "humidity" for a weather service, or "next song" for a music service. Central devices can connect to these peripheral devices, read the services and characteristics, and move on. When no communication is happening, BLE devices operate in a low-power sleep state. Because each data transaction is small and fast, BLE devices spend most of their time in a sleep state and therefore consume very little energy.

PUBLISH AND SUBSCRIBE

Bluetooth LE follows a publish-subscribe model because of a mechanism called *notify*. You can enable the notify feature on a characteristic and have other devices subscribe to notifications for that characteristic. When such a characteristic is updated, a notification about the change is automatically sent to subscribers of the characteristic. This is usually referred to as the *publish-subscribe model*. In Chapter 4, "Exploring Motion Sensors on the Arduino 101," you'll see how to use notify to listen to changes from an accelerometer.

SERVICES, CHARACTERISTICS, AND UUIDS

As mentioned earlier, a BLE peripheral organizes information into services and characteristics. A service can have different characteristics. Depending on the BLE radio you are using, you can create your own services or employ a standard use case. You'll find a full list of adopted BLE services here: *https://www .bluetooth.com/specifications/gatt/services*. For example, Battery Service has an assigned 16-bit UUID of 0x180F and has one mandatory Battery Level characteristic, which has 0x2A19 as its UUID. You'll find a full list of adopted BLE characteristics here: *https://www.bluetooth.com/specifications/gatt/ characteristics*.

Services are identified by unique numbers known as universally unique identifiers (UUIDs). Standard services have a 16-bit UUID, and custom services have a 128-bit UUID. A *characteristic*, on the other hand, is a unit of datum for a given service. It can be up to 20 bytes long. This is a key constraint in designing services. In the following example, you will see how to define a service and a characteristic to light up an LED using Bluetooth LE.

Once you establish a service and a characteristic on a peripheral device, you can program central devices to either read data from the peripheral, write data to certain characteristics on a peripheral, or subscribe to updates on certain characteristics on the peripheral (notify and indicate).

HARDWARE SUPPORT

The experiments and projects in this book require your computer to have hardware and an OS that supports Bluetooth LE. Most new computers and operating systems are BLE enabled, and if you have one that is, you should be all set. If you are on an older device, you can pick up a USB Bluetooth LE adapter for a few bucks on Amazon.

> **NOTE** Windows 7 and older versions of Windows do not support BLE out of the box. If you are a Windows user, I recommend you use Windows 8.1 or Windows 10.

DIAGNOSTIC TOOLS

Punch Through Design's LightBlue for macOS and LightBlue Explorer for iOS are handy diagnostic tools to test your BLE devices. If you are using an Android or Windows device, you can use nRF Connect from Nordic. You can use these tools to

* Discover peripherals

* Connect to a peripheral

* Discover the services of a peripheral

* Discover characteristics of a service

* Read the value of a characteristic

* Subscribe to a characteristic's value

* Write the value of a characteristic

For example, you can connect to your Arduino 101, read values from the Arduino, or write values to it to turn on an LED. Visit these sites to download these tools:

* LightBlue (macOS)—*https://itunes.apple.com/us/app/lightblue/id639944780?mt=12*

FIGURE 3.1: Discovering peripherals, services, and characteristics on LightBlue for macOS

* LightBlue Explorer (iOS)—*https://itunes.apple.com/us/app/lightblue-explorer-bluetooth-low-energy/id557428110?mt=8*

FIGURE 3.2: Discovering peripherals, services, and characteristics on LightBlue for iOS

* nRF Connect for Mobile (Android)—*https://play.google.com/ store/apps/details?id=no.nordicsemi.android.mcp&hl=en*

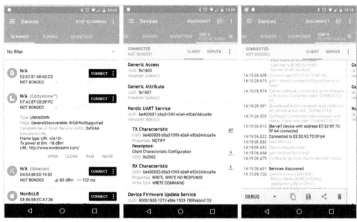

FIGURE 3.3: Discovering peripherals, services, and characteristics on nRF Connect for Mobile for Android

* nRF Connect for Desktop (Windows)—*https://www .nordicsemi.com/eng/Products/Bluetooth-low-energy/ nRF-Connect-for-desktop*, on the Downloads tab

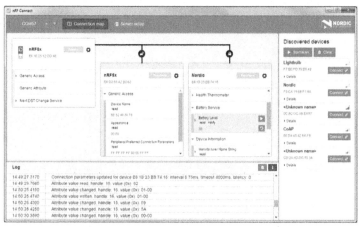

FIGURE 3.4: Discovering peripherals, services, and characteristics on nRF Connect for Desktop for Windows

INTRODUCING THE WEB BLUETOOTH API

The Web Bluetooth API is a new and experimental API that was recently proposed. It is a nonstandard API with limited browser support at this time. But it's very interesting because it allows the browser to connect to devices within 100 meters and read, write, or subscribe to them. As of this writing, it is only supported in Chrome 56 and above. Traditionally, for production websites, it is a better to use a server to communicate with your 101 via Bluetooth LE using libraries like Noble or Bleno. However, the Web Bluetooth API doesn't require you to run a web server, because it directly connects with a web browser. So for the sake of simplicity, we will be using the Web Bluetooth API in this book.

The Web Bluetooth API and libraries like Noble and Bleno open up a lot of possibilities. With this API, you can create user interfaces (UIs), make rich visuals, or design complex use cases for your projects using HTML, CSS, and JavaScript.

For example, you can create a UI for a song composer in the browser and use it to control the Arduino 101 to play different notes and light up LED strips. In addition, you can make use of the motion sensors that the 101 has to control the colors or other visual elements on a web page. Moreover, JavaScript has many libraries that can perform speech recognition, obtain weather information, search for GIFs, and so forth. You can even use the

101 to make your own smart speaker like the Amazon Echo or Google Home. The possibilities are endless, but they all start with establishing a connection between your browser and the 101. In the following chapters, you will learn how to use the Web Bluetooth API to create some fun projects.

> **WARNING** Web Bluetooth API is currently supported in Chrome OS, Android, Mac, Linux, and Windows. It is not currently supported on iOS. Check the browser and platform implementation status here: *https://github.com/WebBluetoothCG/ web-bluetooth/blob/master/implementation-status.md.*

Additional Reading

You can read more about the Web Bluetooth API on the documentation page at *https://developer.mozilla.org/en-US/docs/ Web/API/Web_Bluetooth_API.*

There are a lot of examples to get started with the Web Bluetooth API here: *https://googlechrome.github.io/samples/ web-bluetooth/.*

I also recommend looking at the Noble and Bleno libraries by Sandeep Mistry to create your central and peripheral modules using NodeJS.

Noble: *https://github.com/sandeepmistry/noble*

Bleno: *https://github.com/sandeepmistry/bleno*

BUILDING A SIMPLE WEB PAGE

This is a good point to take a small detour and examine the basics of building a front-end web app using JavaScript, HTML, and

CSS, since we will be using these technologies in the upcoming examples. If you are familiar with the fundamentals of web development, you can skip this section and proceed to the next exercise, which toggles an LED through a web browser.

HTML

Hypertext Markup Language (HTML) is the code that we write to structure our web documents. It uses tags to enclose different types of content so the browser knows how to display them. For example, a paragraph is often enclosed within the `<p>` tag. See the following example:

THE `index.html` **FILE**

```html
<html>
  <head><title>This is my web document</title></head>
  <body>
    <section>
      <h1>This is the heading of a section</h1>
      <p>This is a paragraph within a section in our document.</p>
    </section>
  </body>
</html>
```

CSS

Cascading Stylesheets (CSS) is the code that we write to style our website. It is used to define the look and feel and the visual layout of a website. For the previous example, we can define some styles with CSS as follows:

THE `styles.css` **FILE**

```css
/* Style to add a margin to our section */
section {
  margin: 20px;
}

/* Style to change the font size and color of our heading and
paragraph */
```

```
h1 {
  font-size: 20px;
  color: blue;
}

p {
  font-size: 10px;
  color: black;
}
```

JavaScript

JavaScript is the programming language of the web. This is what we write to add interactive features, submit data from forms, talk to our server and database, and determine the behavior of different components on our page. Consider the following example, which changes the text of our section paragraph when it runs.

THE index.js **FILE**

```
var myParagraph = document.querySelector('p');
myParagraph.textContent = 'Hello world!';
```

The browser will open and display our index.html file. We need to do one more thing: link our styles.css and index.js files to our HTML file so they can be imported and processed by the browser accordingly. For this, we need to add a <link> tag and a <script> tag to our HTML document.

THE index.js **FILE WITH CSS AND JAVASCRIPT**

```
<html>
  <head>
    <title>This is my web document</title>
    <!-- This assumes that our index.js and styles.css files are in
the same folder as this file -->
    <link rel="stylesheet" href="styles.css">
  </head>
  <body>
    <section>
      <h1>This is the heading of a section</h1>
      <p>This is a paragraph within a section in our document.</p>
    </section>
```

```
    <script src="index.js"></script>
  </body>
</html>
```

Now that you have created and saved these files in the same folder, you can go ahead and open the index.html file in your favorite web browser to see your web page.

Additional Reading

This was just a brief introduction to web development and how HTML, CSS, and JavaScript can be used to create interactive websites. If you are curious to learn more, I recommend that you go through MDN's excellent free online tutorials on web fundamentals (*https://developer.mozilla.org/en-US/docs/Learn*).

Now that we've talked about Bluetooth LE, the Web Bluetooth API, and web development basics, let's rebuild our blinking LED project from Chapter 2 with the power of the web!

PROJECT: TOGGLE AN LED VIA A CHROME BROWSER

In this project, you are going to blink an LED via Bluetooth LE on the 101 and a BLE-capable computer. The idea is to connect the 101 to a Chrome web browser on a computer through BLE, and then use an input box on a web page to toggle an LED on and off.

You'll need the following (Figure 3.5):

* An Arduino 101 (with a USB cable)

* A battery jack

* A 9V battery

* An LED

* A BLE-enabled computer with the Google Chrome web browser installed on it

FIGURE 3.5: What you'll need for our "Toggle an LED via a Chrome Browser" project

Circuit Assembly

The circuit will be exactly same as the one you used in our "Blinking an LED" project from Chapter 2. You will connect an LED between the Arduino 101's pin number 13 and GND pin. Remember to insert the LED's longer pin to the Arduino 101's pin number 13 (see Figure 3.6).

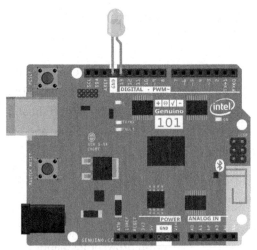

FIGURE 3.6: Circuit and component diagram of our "Toggle an LED via a Chrome Browser" project

System Diagram

Figure 3.7 highlights the chain of events that happen during the exchange of data over BLE.

1. Browser will requestDevice() with a specified Service UUID and a name
2. If a device is found, connect to it
3. Get service with a specified Service UUID
4. Get the first characteristic
5. Write value 1 or 0 to Arduino 101

FIGURE 3.7: System diagram of our "Toggle an LED via a Chrome Browser" project

Code

You can find the complete code here:

https://github.com/yining1023/Jumpstarting-the-Arduino-101

In this project, you need two programs: one that runs on the Arduino and another that runs in your browser.

The Arduino sketch will create a BLE peripheral with a name, a UUID, a service, and a characteristic. It will also tell the 101 to turn the LED at pin 13 on or off when a central device writes a value to its characteristic.

The other program runs in your browser. It will create a web page with HTML and CSS and will have a Connect button, an input box, and a Send button. When you click the Connect button, JavaScript code will use the Web Bluetooth API and act as a central device. The browser will then search for your Arduino by a particular name and UUID. When you click the Send button, the JavaScript code will write the value you put in the input box to your 101 and turn on the LED.

Arduino Sketch

For this example you will be using a slightly modified version of a built-in example provided by the CurieBLE library. Choose File > CurieBLE > Peripheral > CallbackLED and open the example sketch as shown in Figures 3.8 and 3.9.

This time you will make a small modification to the following line:

```
if (switchChar.value())
```

Change it to this:

```
if (switchChar.value() == 1)
```

FIGURE 3.8: Opening the CallbackLED example

```
● ● ●                    CallbackLED | Arduino 1.6.9
⊘⊘ ▣ ▾ ▯                                                    ▣
CallbackLED                                                 ▾
1 /*
2  * Copyright (c) 2016 Intel Corporation.  All rights reserved.
3  * See the bottom of this file for the license terms.
4  */
5
6 #include <CurieBLE.h>
7
8 const int ledPin = 13; // set ledPin to use on-board LED
9
10 BLEService ledService("19B10000-E8F2-537E-4F6C-D104768A1214"); // create service
11
12 // create switch characteristic and allow remote device to read and write
13 BLECharCharacteristic switchChar("19B10001-E8F2-537E-4F6C-D104768A1214", BLERead | BLEWrite);
14
15 void setup() {
16   Serial.begin(9600);
17   pinMode(ledPin, OUTPUT); // use the LED on pin 13 as an output
18
19   // begin initialization
20   BLE.begin();
21
22   // set the local name peripheral advertises
23   BLE.setLocalName("LEDCB");
24   // set the UUID for the service this peripheral advertises
25   BLE.setAdvertisedService(ledService);
26
27   // add the characteristic to the service
28   ledService.addCharacteristic(switchChar);
29
30   // add service
31   BLE.addService(ledService);
32
33   // assign event handlers for connected, disconnected to peripheral
34   BLE.setEventHandler(BLEConnected, blePeripheralConnectHandler);
35   BLE.setEventHandler(BLEDisconnected, blePeripheralDisconnectHandler);
36
                                                      ●
                          Arduino/Genuino 101 on /dev/cu.usbmodem1423
```

FIGURE 3.9: CallbackLED example code

switchChar is the name of a characteristic we created for a
service, ledService, that will be running on our peripheral 101,
as you will see later. The change in the if block means that the
code under the if block will run only when switchChar's value is
1. Since you are turning on the LED under that if block, the LED
will turn on when switchChar.value() is 1. The example from the
CurieBLE library is telling 101 to turn on the LED only if there is
a value, no matter what that value is. But you also want to have
the ability to turn off the LED for certain values. With this change,
if the characteristic's value is 1, then the 101 will turn on the LED.
For no value or all other values, the LED will be turned off.

Let's start by importing the CurieBLE library. Then, define a constant, `ledPin`, and assign it a value of 13 to represent pin number 13.

Next, define a BLE service and a characteristic. Name the service `ledService` and the characteristic `switchChar`, and create them with a UUID. There are several ways to generate UUIDs. You can generate a UUID on *https://www.uuidgenerator.net/*. The service UUID and the local name are used for discovery, as you will see later.

You are creating a BLE `Char` characteristic, which means that your characteristic's value will have a data type of character.

```
/*
 * Copyright © 2016 Intel Corporation.  All rights reserved.
 * See the bottom of this file for the license terms.
 */

#include <CurieBLE.h>
const int ledPin = 13; // set ledPin to use on-board LED

// create service
BLEService ledService("19B10000-E8F2-537E-4F6C-D104768A1214");

// create switch characteristic and allow remote device to read and
write
BLECharCharacteristic switchChar("19B10001-E8F2-537E-4F6C-
D104768A1214", BLERead | BLEWrite);
```

In your `setup()` function, after setting pin 13 as OUTPUT, initialize the BLE library instance. Then set a local name and advertise a service UUID for your Arduino 101 peripheral device. The local name and service UUID will be advertised and used by central devices to connect to this peripheral.

Next, add your `switchChar` characteristic to your `ledService` and list your `ledService` on the peripheral device.

The next step is to set up some event handlers for different events. Here, the `blePeripheralConnectHandler` function will be called on the `BLEConnected` event. Similarly, the `blePeripheralDisconnectHandler` will be called when the `BLEDisconnected` event is fired.

Then, assign an event handler, `switchCharacteristicWritten`, to the `BLEWritten` event.

Before you start advertising your peripheral to potential central devices, assign an initial value of 0 to your switch characteristic.

You will also define your event handler functions: `blePeripheralConnectHandler`, `blePeripheralDisconnectHandler`, and `switchCharacteristicWritten`.

```
void setup() {
  Serial.begin(9600);
  pinMode(ledPin, OUTPUT); // use the LED on pin 13 as an output

  // begin initialization
  BLE.begin();

  // set the local name peripheral advertises
  BLE.setLocalName("LEDCB");
  // set the UUID for the service this peripheral advertises
  BLE.setAdvertisedService(ledService);

  // add the characteristic to the service
  ledService.addCharacteristic(switchChar);

  // add service
  BLE.addService(ledService);

  // assign event handlers for connected, disconnected to peripheral
  BLE.setEventHandler(BLEConnected, blePeripheralConnectHandler);
  BLE.setEventHandler(BLEDisconnected,
blePeripheralDisconnectHandler);

  // assign event handlers for characteristic
  switchChar.setEventHandler(BLEWritten,
switchCharacteristicWritten);
  // set an initial value for the characteristic
  switchChar.setValue(0);

  // start advertising
  BLE.advertise();
```

```
Serial.println(("Bluetooth device active, waiting for
connections..."));
}
```

In the `loop()` function, you poll for BLE events. Since `loop()`
is called continuously, you will poll the radio for BLE events periodically throughout the life of the sketch.

```
void loop() {
  // poll for BLE events
  BLE.poll();
}
```

Now you'll define your connect and disconnect event handlers that I mentioned earlier. In both the event handlers, you
are printing a log line with the address of the central device to
the `Serial`.

```
void blePeripheralConnectHandler(BLEDevice central) {
  // central connected event handler
  Serial.print("Connected event, central: ");
  Serial.println(central.address());
}

void blePeripheralDisconnectHandler(BLEDevice central) {
  // central disconnected event handler
  Serial.print("Disconnected event, central: ");
  Serial.println(central.address());
}
```

Finally, you'll define your `switchCharacteristicWritten` event
handler, which will mostly be used for handling the LED logic that
you want. It is here that you compare the newly written value of
the `switch` characteristic. If it is 1, you turn on the LED; otherwise,
you turn it off.

```
void switchCharacteristicWritten(BLEDevice central,
BLECharacteristic characteristic) {
  // central wrote new value to characteristic, update LED
  Serial.print("Characteristic event, written: ");
  if (switchChar.value() == 1) {
    Serial.println("LED on");
    digitalWrite(ledPin, HIGH);
```

```
  } else {
    Serial.println("LED off");
    digitalWrite(ledPin, LOW);
  }
}
```

Run the Arduino Code

Next, in the Arduino IDE, choose the correct board and port
as described in Chapter 1, "What Is Arduino?" Select Tools >
Boards > Arduino/Genuino 101, and then choose Tools > Port.
Click the Verify button as in the previous example. Once the
sketch has compiled successfully, click Upload. If you get a "Done
uploading" message, that means the code is uploaded to the 101
successfully!

FIGURE 3.10: The "Done uploading" message in the Arduino IDE

Test the Arduino Code

At this point, because you have two devices in this project, it's a good practice to test the Arduino code first before you move on to the next step. In this way, if you make sure the Arduino code works as expected, and in the future you cannot get the result you want, you will know which program to debug.

You can use the LightBlue (macOS and iOS) or nRF Connect (Android and Windows) diagnostic tool to test your Arduino code. Turn on the Bluetooth on your computer, and open the LightBlue app. If you see a peripheral named LEDCB, it means your 101 is advertising itself correctly. Select this peripheral, along with the service and characteristic UUIDs that you defined. Then, type **0x01** in the Write HEX input box. 0x01 is the hexadecimal value for a decimal 1. If the LED on the Arduino lights up, it means that your Arduino code works well, and you are good to proceed to the next step!

FIGURE 3.11: Using LightBlue on macOS to connect to the 101 and write 0x01 to it

Now you'll write some JavaScript code to build a web page to talk to your 101. If the LED doesn't light up, check the "Trouble-shooting Tips" section at the end of this project.

JavaScript Code

At the beginning of this chapter, we talked about how to build a simple web page. For this project, you'll write three files: `index.html`, `style.css`, and `main.js`. You can create a folder and use any editor to write these files and save them in that folder.

> **NOTE** Sublime Text (*https://www.sublimetext.com/*) and Atom (*https://atom.io/*) are two of my favorite editors, because they have thousands of great and versatile open source plugins available.

index.html

As mentioned in the "Building a Simple Web Page" section earlier, you'll create an HTML document. Then you'll add a heading with some instructions in it, and create an input box with the ID `input-box` and a Connect button. Once you click the Connect button, the `connectTo101()` function from your JavaScript file will be called.

You'll also create a "Write to 101" button, which when clicked will call the `writeTo101()` function defined in your `main.js` file.

```
<!doctype html>
<html lang="en">
  <head>
    <meta charset="utf-8">
    <title>Arduino 101 BLE LED</title>
  </head>

  <body>
    <h1>Write 1 or 0 to turn the LED on or off</h1>
    <button onclick="connectTo101()">Connect</button>
    Write Value: <input type="text" name="writeValue"
id="input-box">
```

```
    <button onclick="writeTo101()">Write To 101</button>
    <script src="main.js"></script>
  </body>
</html>
```

For this project, we'll skip styling with CSS. I encourage you to read through some of the styling fundamentals from Mozilla Developer Network (MDN) (*https://developer.mozilla.org/en-US/ docs/Learn/Getting_started_with_the_web/CSS_basics*) and add some CSS to improve the appearance of your document.

Next, you'll create the `main.js` file for your JavaScript code.

main.js

Let's start off by defining some constants. `serviceUuid` and `name` will allow this central device (your browser) to connect to the correct service on your peripheral 101 device with these two identifiers.

```
const serviceUuid ="19b10000-e8f2-537e-4f6c-d104768a1214";
const name = 'LEDCB';

var ledCharacteristic;
```

Next, define the `connectTo101()` function, which will be called when you click the Connect button in the UI. In this function, you first create an `options` object, which defines filters to search for your BLE peripheral device. Then you use the Web Bluetooth API to `requestDevice` and find your peripheral device with given filters.

The `navigator.bluetooth.requestDevice()` function returns a JavaScript `Promise` object. A `Promise` is an object that is often used for asynchronous communications with JavaScript, such as talking to a server. Once you receive a response, the callback function in the `Promise`'s `then` handler is executed. If there was an error, the callback function in the `catch` handler is executed. You can learn more about `Promises` by checking out this excellent explanation by Google: *https://developers.google.com/web/ fundamentals/getting-started/primers/promises.*

Next, you connect to this device and then get the service with a given UUID from this device. Once you get the service, you get its characteristics using `service.getCharacteristics()`.

Now that you have the characteristics, you assign the first one in the `characteristics` array to your `ledCharacteristic` variable.

```javascript
function connectTo101() {
  let options = {
    filters: [{
      services: [serviceUuid],
      name: name
    }]
  }

  console.log('Requesting Bluetooth Device...');
  navigator.bluetooth.requestDevice(options)
    .then(device => {
      console.log('Got device', device.name);
      return device.gatt.connect();
    })
    .then(server => {
      console.log('Getting Service...');
      return server.getPrimaryService(serviceUuid);
    })
    .then(service => {
      console.log('Getting Characteristics...');
      return service.getCharacteristics();
    })
    .then(characteristics => {
      console.log('Got LED Characteristic');
      ledCharacteristic = characteristics[0];
    })
    .catch(error => {
      console.log('Argh! ' + error);
    });
}
```

Lastly, you define your `writeTo101()` function. In this function, you first select the value of your input with an ID of `input-box` and get its value. Then you convert the value to a buffer array of 8-bit unsigned integers. You can then use the `writeValue` function from the Web Bluetooth API to write this buffer to the `ledCharacteristic` that was selected in the `connectTo101()` function.

You've now written a characteristic value to your peripheral 101 device. The Arduino sketch you wrote earlier will handle the rest.

```
function writeTo101() {
  let inputValue = document.getElementById('input-box').value;

  let bufferToSend = Uint8Array.of(inputValue);
  ledCharacteristic.writeValue(bufferToSend);
  console.log('Writing '+ inputValue + ' to led Characteristic...');
}
```

Run the JavaScript Code

Now if you open the `index.html` file in the Chrome browser, you should see something like the screen shown in Figure 3.12.

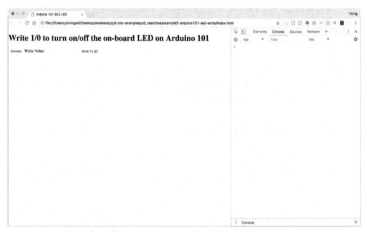

FIGURE 3.12: Rendered `index.html` in the Chrome browser

Final Result

Before you test the final result, you must do two things:

* Power the Arduino 101 with the code you just uploaded using a 9V battery and a battery jack like in Figure 3.13.

* On your computer, make sure Bluetooth is turned on.

FIGURE 3.13: The Arduino 101 with a 9V battery and a battery jack

FIGURE 3.14: Final setup

Now that your code is done and your hardware is all set up, let's test our final result.

Open the Chrome inspector by pressing Cmd-Shift-I on a Mac or Ctrl-Shift-I on Windows or Linux. Then navigate to the Console tab on the inspector. Click the Connect button on the web page. A dialog should pop up asking for permission to pair with a peripheral. Select LEDCB, as shown in Figure 3.15, and click Pair.

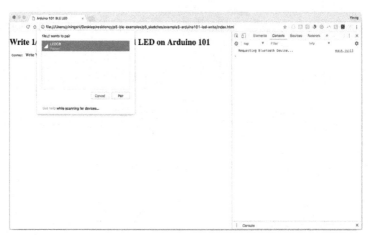

FIGURE 3.15: Connecting the Arduino 101 in the browser

You should see the logs you wrote in your `connectTo101()` function from `main.js` in the console after you click Connect.

Now, enter **1** or **0** in the text box and click the Write To 101 button to send either 1 or 0 to your 101 board. You should then see the LED turn on or off.

You can also view a video demo for the project here: *https://youtu.be/pcE36zL3A38.*

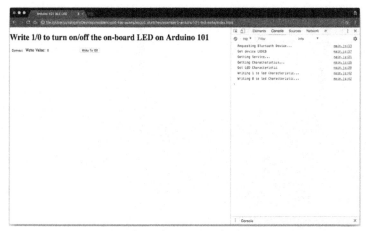

FIGURE 3.16: Writing 0 to the Arduino 101 in the browser

FIGURE 3.17: The final result for the "Toggle an LED via a Chrome browser" project (1)

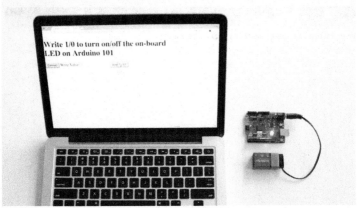

FIGURE 3.18: The final result for the "Toggle an LED via a Chrome browser" project (2). After you enter 1 in the input box and click the Write To 101 button, the LED on the Arduino 101 will light up.

FURTHER EXPLORATION

Congratulations! You finished the "Toggle an LED via a Chrome Browser" project. For your next step, there are a lot of interesting possibilities. You can use a color picker in the browser to pick a color and light up an RGB LED on the Arduino with the color you picked. You can use LED strips or lamps, too. Or you can use an Android phone with the Web Bluetooth API to control the Arduino 101 instead of using the desktop app.

Troubleshooting Tips

* Don't forget to turn on the Bluetooth on your computer.

* Make sure Arduino code and JavaScript code have the same local name and Service UUID for the Arduino 101.

* After uploading the code in the Arduino IDE, wait for five seconds to open the serial window.

Exploring Motion Sensors on the Arduino 101

From activity tracking monitors to optical image stabilization in cameras, the applications of motion sensing are endless. In this chapter, you will be exploring how the Arduino 101 can be used to detect and measure motion in a three-dimensional space.

INERTIAL MEASUREMENT UNIT

The Arduino 101 has an onboard six-axis motion sensor, also known as an inertial measurement unit (IMU). The term *inertial measurement unit* usually refers to a device that measures the specific force on a body by using various sensors, such as an accelerometer or a gyroscope. The Arduino 101 has a three-axis accelerometer and a three-axis gyroscope.

The three-axis accelerometer measures the board's proper acceleration along three axes—X, Y, and Z—in a three-dimensional space. This means that when the accelerometer is at rest on the surface of Earth, it will always measure an acceleration of –9.8 m/s^2 along the Z-axis—the acceleration due to Earth's gravity. The gyroscope measures the angular motion or the rotational rate of an object. It can tell you how much the board is tilted left to right, back to front, and side to side.

The IMU can specifically measure the roll, pitch, and yaw of a given body. Figure 4.1 sums up the difference between the three axes.

FIGURE 4.1: Roll, pitch, and yaw of a given body

In this chapter, we'll explore the various things possible with Curie's IMU. The Curie MU library (*https://www.arduino.cc/en/Reference/CurieIMU*) provides several examples that will help you understand the 101's IMU data. You can find them under File > Examples > CurieIMU in the Arduino IDE (Figure 4.2) if you have installed the Intel Curie boards, as explained in Chapter 2, "Getting Started." The built-in examples include not only reading the accelerometer and gyroscope data, but also counting steps and detecting free-fall motion, shocks, taps, and double taps.

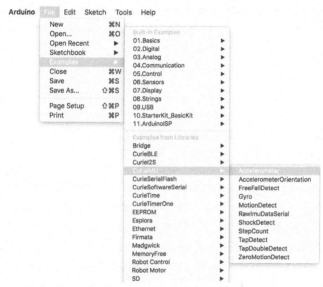

FIGURE 4.2: Examples from CurieIMU in the Arduino IDE

I recommend going to *https://www.arduino.cc/en/Tutorial/Genuino101CurieIMUOrientationVisualiser* and running the Arduino/Genuino 101 CurieIMU Orientation Visualiser tutorial. Don't forget to uncomment and edit the port number in the code based on your operating system.

In the following project, you will be using the p5.js JavaScript library to create a visualization about the motion sensor data you get from the 101 through Bluetooth Low Energy (BLE). Using this library, you can create graphics and animations in the browser as easily as you can with Processing. Just like the Arduino language, the p5.js project is entirely modeled after the Processing language and is maintained by the Processing Foundation.

> **NOTE** To learn more about p5.js and Processing, visit the Getting Started guide (*https://p5js.org/get-started/*) and Reference page (*https://p5js.org/reference/*) for p5.js. *Make: Getting Started with p5.js* is a great book by Lauren McCarthy, Casey Reas, and Ben Fry (Maker Media, 2015) that you can refer to learn more about p5.js.

PROJECT: CONTROLLING A WEB PAGE WITH AN INTERACTIVE TOY

In this example, you will detect taps and double taps with the IMU sensor on the Arduino 101. In Chapter 3, "Exploring Bluetooth LE on the Arduino 101," we discussed how to write data to the 101 over Bluetooth. Our next example will mostly focus on reading IMU data from the 101 into the browser. You will also create an interactive web page using p5.js and use the 101 to interact with the web page.

Optionally, you will be packing your 101 nicely inside a soft toy to build a playful experience around it.

System Diagram

Figure 4.3 shows the system diagram for this project.

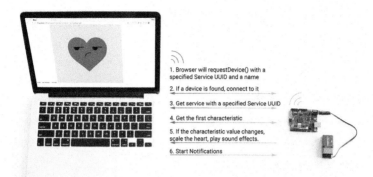

1. Browser will requestDevice() with a specified Service UUID and a name

2. If a device is found, connect to it

3. Get service with a specified Service UUID

4. Get the first characteristic

5. If the characteristic value changes, scale the heart, play sound effects.

6. Start Notifications

FIGURE 4.3: System diagram for the "Controlling a Web Page with an Interactive Toy" project

What You'll Need

You'll need the following for this project (Figure 4.4):

* An Arduino 101 (with a USB cable)

* A 9V battery

* A battery jack

FIGURE 4.4: What you'll need for the "Controlling a Web Page with an Interactive Toy" project

* A BLE-enabled computer with the Google Chrome web browser installed on it

* Optionally, a small stuffed toy or an enclosure of your choice

Circuit Assembly

Since you are using the Arduino 101's onboard IMU sensor, there is no need for any external circuit. But you do need a battery and battery jack, as shown in Figure 4.4.

Code

You'll find the complete code here:

https://github.com/yining1023/Jumpstarting-the-Arduino-101

Arduino Sketch

As in our previous project, you'll start off by importing the required Curie libraries and defining a service called `tapService`. You'll also create a characteristic called `tapCharacteristic` of type `unsigned char`.

You'll also define a variable named `tapValue` and initialize it with a value of 0. You'll use this variable to keep track of the number of times a user has tapped the sensor. For our purposes, it will be 0 (no taps), 1 (a single tap), or 2 (a double tap).

```
#include <CurieBLE.h>
#include "CurieIMU.h"

BLEPeripheral blePeripheral;

// create service
BLEService tapService = BLEService("19B10000-E8F2-537E-
4F6C-D104768A1214");

// create tap characteristic and allow remote device to read and write
BLEUnsignedCharCharacteristic tapCharacteristic =
BLEUnsignedCharCharacteristic("5667f3b1-d6a2-4fb2-a917-4bee580a9c84",
```

```
BLERead | BLENotify);

int tapValue = 0;
```

In the setup function, you'll set up your service and characteristic just as you did in the previous project. You'll then initialize the CurieIMU library and attach an interrupt action called
eventCallback. This function, which you'll define later, will be
called upon whenever any interrupt is detected by the IMU. For
our project, we mostly care about two interrupts: CURIE_IMU_TAP
and CURIE_IMU_DOUBLE_TAP.

> NOTE For a list of all the predefined interrupts, refer to
> the API reference at https://www.arduino.cc/en/Reference/
> CurieIMUinterrupts.

Next you'll set up the accelerometer range and the detection
thresholds for the interrupts. Think of this as setting the sensitivity of the IMU to fit your needs. Finally, you'll set the detection
duration for the double-tap event and enable the two interrupts.

```
void setup() {
  Serial.begin(9600); // initialize Serial communication

  // set the local name peripheral advertises
  blePeripheral.setLocalName("CurieTap");
  // set the UUID for the service this peripheral advertises
  blePeripheral.setAdvertisedServiceUuid(tapService.uuid());

  // add service and characteristic
  blePeripheral.addAttribute(tapService);
  blePeripheral.addAttribute(tapCharacteristic);

  blePeripheral.begin();

  tapCharacteristic.setValue(tapValue);
```

```
    Serial.println(("Bluetooth device active, waiting for
connections..."));

    // Initialize the IMU
    CurieIMU.begin();
    CurieIMU.attachInterrupt(eventCallback);

    // Increase Accelerometer range to allow detection of stronger
    // (< 4g)
    CurieIMU.setAccelerometerRange(4);

    // Reduce threshold to allow detection of weaker taps (>= 750mg)
    CurieIMU.setDetectionThreshold(CURIE_IMU_TAP, 750); // (750mg)

    // Reduce threshold to allow detection of weaker taps (>= 750mg)
    CurieIMU.setDetectionThreshold(CURIE_IMU_DOUBLE_TAP, 750);

    // Set the quiet time window for 2 taps to be registered as a
    // double-tap (Gap time between taps <= 1000 milliseconds)
    CurieIMU.setDetectionDuration(CURIE_IMU_DOUBLE_TAP, 1000);

    // Enable Tap detection
    CurieIMU.interrupts(CURIE_IMU_TAP);

    // Enable Double-Tap detection
    CurieIMU.interrupts(CURIE_IMU_DOUBLE_TAP);

    Serial.println("IMU initialization complete, waiting for
events...");
}
```

As in our previous project, in your `loop` function you continuously poll the BLE radio for events. As mentioned earlier, you'll define an `eventCallback`, which outlines the logic for the steps to take once the 101 detects a tap or a double-tap interrupt. In this case, you want to set the `tapValue` and write that value to the `tapCharacteristic` so that any connected central device can read the characteristic.

```
void loop() {
    // Tell the Bluetooth radio to poll events
    blePeripheral.poll();
```

```
}
static void eventCallback()
{
  if (CurieIMU.getInterruptStatus(CURIE_IMU_DOUBLE_TAP)) {
    Serial.println("DOUBLE Tap");
    tapValue = 2;
    tapCharacteristic.setValue(tapValue);
  } else if (CurieIMU.getInterruptStatus(CURIE_IMU_TAP)) {
    Serial.println("SINGLE Tap");
    tapValue = 1;
    tapCharacteristic.setValue(tapValue);
  }
}
```

Testing the Arduino Sketch

Verify and upload your sketch with the USB cable to your 101.
Then open the serial monitor and tap or double tap your 101 to
find the sweet spot for detection. You can see the results on the
serial monitor because you are printing some helpful statements
in your eventCallback. You can configure the sensitivity of your
taps by playing around with the detection threshold and detection duration in the previous code.

Before we move on to the JavaScript code, it is a good idea
to test the behavior of the BLE peripheral of the 101. You can use
the LightBlue app (for Mac) or the nRF app (for Windows/Linux)
to do this, just as you did in Chapter 3.

Web App Code

To enclose my Arduino, I chose a small, heart-shaped soft toy that
I had lying around. If you live near an Ikea store, you can buy it for
99 cents! For this project I also decided to have a heart-shaped
emoji on the website react to the taps on the soft toy with the
word "Ouch!" Let's walk through the web code.

index.html

```
<!doctype html>
<html lang="en">
  <head>
    <meta charset="utf-8">
    <title>BLE</title>
    <!-- import p5.js and p5.sound libraries -->
    <script src="https://cdnjs.cloudflare.com/ajax/libs/p5.js/0.5.4/
p5.min.js"></script>
    <script src="https://cdnjs.cloudflare.com/ajax/libs/p5.js/0.5.4/
addons/p5.sound.min.js"></script>
  </head>

  <body>
    <button onClick="connect()">Connect</button>
    <button onClick="startNotify()">Start Notifications</button>
    <button onClick="disconnect()">Disconnect</button>
    <script src="main.js"></script>
  </body>
</html>
```

main.js

As always, you'll start by defining some required variables:

```
const serviceUuid = "19b10000-e8f2-537e-4f6c-d104768a1214";
const name = 'CurieTap';

var bluetoothDevice;
var tapCharacteristic;
var tapValue;

// the 2 versions of the heart emoji
var heartImage, cryheartImage;

var ouch, ouchouch; // used for 2 versions of sound files
var scaleValue = 4; // used to scale the image
```

The connect function is going to be similar to the previous example, but with one small change:

```
function connect() {
  let options = {
    filters: [{
```

```
    services: [serviceUuid],
    name: name
  }]}

console.log('Requesting Bluetooth Device...');

navigator.bluetooth.requestDevice(options)
.then(device => {
  bluetoothDevice = device; // save a copy
  console.log('Got device', device.name);
  return device.gatt.connect();
})
.then(server => {
  console.log('Getting Service...');
  return server.getPrimaryService(serviceUuid);
})
.then(service => {
  console.log('Getting Characteristics...');
  // Get all characteristics.
  return service.getCharacteristics();
})
```

This time around you will add an event listener to the
characteristicvaluechanged event. You know from your Arduino
sketch that the tapCharacteristic value changes on a tap or a
double-tap detection. Here you're setting up a handleTap function
that will be called every time the characteristic value is changed.

```
.then(characteristics => {
  console.log('Got Characteristics');
  tapCharacteristic = characteristics[0];
  // add an event listener for changes in the characteristic
  tapCharacteristic.addEventListener(
    'characteristicvaluechanged',
    handleTap
  );
})
.catch(error => {
  console.log('Argh! ' + error);
});
}
```

You also want to listen to notifications from the peripheral. Think of this step as subscribing to change events on the characteristic.

```
function startNotify() {
  tapCharacteristic.startNotifications();
}
```

The `handleTap` function, which is attached to your event handler, will then play a sound and scale the heart image depending on the number of taps, as you'll see next.

```
function handleTap(event)
{
  tapValue = event.target.value.getUint8(0);
  console.log('tap: ' + tapValue);
  playSound();
  scaleImage();
}
```

`preload` is a function that is called by p5.js. Just like the `setup` and `loop` functions in Arduino, p5.js has `setup` and `draw` functions. It also has a `preload` function that is used to load files asynchronously before your p5 sketch starts (i.e., before the `setup` function). The `setup` function won't be called until everything in the `preload` function is done. In this way, you make sure that the `setup` and `draw` functions have access to the image and sound files.

> NOTE For more information, refer to the p5.js documentation at *https://p5js.org/reference/#/p5/*.

```
function preload() {
  heartImage = loadImage("assets/heart.png");
  cryheartImage = loadImage("assets/cryheart.png");
  ouch = loadSound('assets/ouch.m4a');
  ouchouch = loadSound('assets/ouchouch.m4a');
}
```

In your p5.js `setup` function, you'll start by creating a `<canvas/>` element that will render your sketch. You'll also center the image (see *https://p5js.org/reference/#/p5/createCanvas*).

```
function setup() {
  createCanvas(800, 700);
  imageMode(CENTER);
}
```

Inside the `draw` function the `image` function is called periodically throughout the life of the sketch to render each frame on your `<canvas/>` element. It is usually called 60 times per second.

The `image` function is then called. It draws the `heartImage` and scales it depending on the `scaleValue` variable.

`scaleValue` will also be used to switch between the drawn images. If a tap is detected, a crying image is drawn; otherwise, a frowning image is drawn.

As you will see later, `scaleValue` is changed every time a tap is detected using the `scaleImage` function. This creates a nice animation effect.

```
function draw() {
  background('#FFE3DF');

  if (scaleValue <= 4) {
    scaleValue = 4;
    // draw the image at a scaled width and height
    image(
      heartImage,
      width / 2,
      height / 2,
      heartImage.width / scaleValue,
      heartImage.height / scaleValue
    );
  } else {
    // if the scale value is greater than 4,
    // which means the image is small
    // then swap the image with a crying image
    // this means the heart
    // will be sad every time it's tapped
    image(
```

```
    cryheartImage,
    width / 2,
    height / 2,
    heartImage.width / scaleValue,
    heartImage.height / scaleValue
  );
  // gradually increase the scaleValue with
  // every draw call
  scaleValue -= 0.2;
  }
}
```

You will now define the `scaleImage` and the `playSound` functions, which are called in `handleTap`. As you can see in the code comments, these functions do different things when either a single or a double tap is detected.

```
function scaleImage() {
  if (tapValue === 1) {
    // for single tap scale down by a factor of 8
    scaleValue = 8;
  } else if (tapValue === 2) {
    // for double tap scale down by a factor of 4
    scaleValue = 10;
  } else {
    // let the default scaleValue be 4
    scaleValue = 4;
  }
}

function playSound() {
  if (tapValue === 1) {
    // for single tap, play the ouch sound file
    ouch.setVolume(0.5);
    ouchouch.stop();
    if (ouch.isPlaying()) ouch.stop();
    ouch.play();
  } else if (tapValue === 2) {
    // for double tap, play the ouchouch sound file
    ouchouch.setVolume(0.5);
    ouch.stop();
    if (ouchouch.isPlaying()) ouchouch.stop();
    ouchouch.play();
  }
}
```

Lastly, you'll define a `disconnect` function, which will be called when you click the Disconnect button.

```
function disconnect() {
  if (bluetoothDevice && bluetoothDevice.gatt) {
    bluetoothDevice.gatt.disconnect();
    console.log('Disconnected');
  }
}
```

RUNNING THE APP

This time, instead of simply opening your `index.html` file in Google Chrome, you are going to serve it from a static HTTP server. This is because you are loading sound and images.

Open your Terminal app (on Mac or Linux) or PowerShell on Windows. On most macOS and Linux machines, Python will be installed by default. If not, you can download and install Python 2.x or 3.x from *https://www.python.org/downloads/*.

Type `cd` and drag the folder in which you have your JavaScript and HTML files into your Terminal window. Alternatively, you can type the full path of your directory after `cd` and press the Enter/Return key. Then type the following command for Python 2.x to start a simple Python HTTP server:

```
python -m SimpleHTTPServer
```

If you are using Python 3x, type the following command instead:

```
python3 -m http.server
```

The `simpleHTTPServer` is by default served on port 8000, so open your Chrome browser and type **localhost:8000** in your address bar to see your web page!

Fabrication (Optional)

Here are some of the photos of the enclosure we chose for our project.

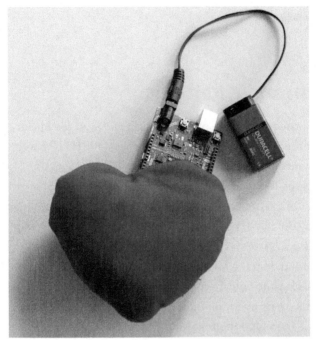

FIGURE 4.5: Put your Arduino and the battery into a stuffed toy.

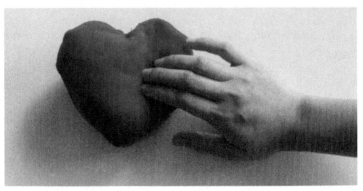

FIGURE 4.6: Tap the stuffed toy.

FINAL RESULTS

The final results are shown in the following figures.

FIGURE 4.7: The web page with the p5.js app

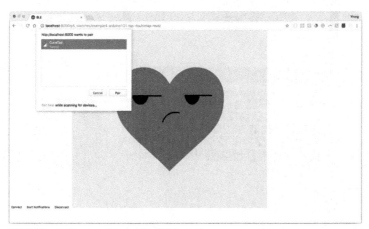

FIGURE 4.8: Connecting to the 101 board

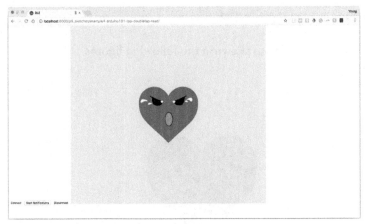

FIGURE 4.9: When a tap is detected

FIGURE 4.10: The final setup

You can also view a video demo for the project here:

https://youtu.be/aS4TPNicFsA

FURTHER EXPLORATION

At this point, you can read any sensor data from the 101 and send it to the browser to create visuals or sounds. Then, publish your website. There are a few options; see *https://developer.mozilla .org/en-US/docs/Learn/Getting_started_with_the_web/Publishing _your_website*. I like to use GitHub pages to publish a simple website. See GitHub's blog post here: *https://help.github.com/ articles/what-is-github-pages/*.

5

Exploring Pattern Matching and Machine Learning on Intel Curie

My favorite thing about the Arduino 101 is that the Curie module comes with a neural network chip built in. I think this is very commendable for an entry-level, affordable board, as it enables you to use machine learning in your projects. The QuarkSE module on the Curie comes with 128 neurons, with 128 bytes of memory per neuron.

WHAT IS MACHINE LEARNING?

A *neural network* is a collection of computational units that can be trained (or can train themselves) to solve problems the way a human brain would. Just like a human brain, a neural network is made up of a number of interconnected subunits called *neurons*. Think of a neuron as a function that takes one or more inputs and produces an output after processing some logic. That logic is what it learns over time, usually through human guidance or examples.

In traditional programming, we write algorithms that tell the machine what to do and how to do it exactly. In other words, the logic is designed by programmers. But with machine learning, the problem is to design an algorithm that will train a machine to look at a series of inputs and their outputs in order to find a pattern or a relationship between them. Once that is done, it will use its training to *guess* an output for a given input.

> **NOTE** I've explained this in very simple terms. The whole process is often a mathematically difficult problem to address.

Thankfully, Curie's pattern-matching engine and related libraries make the whole process very easy. All the machine-learning features are available to us as simple, easy-to-use functions. To end users they are available at the press of a button!

THE CuriePME LIBRARY

The CuriePME library offers a great set of functions that you can use for training the neural network on the Arduino 101. This is exciting because it can take any arbitrary data as input and try to find the pattern no matter what the data is. The data can come from an IMU, a thermometer, light sensors, force sensors, or even a static list of data of your choice, like numbers or letters.

The CuriePME library is mostly used for the following:

* Learning patterns

* Recognizing and classifying patterns

* Storing and retrieving pattern memory as knowledge (requires the SerialFlash library)

Installation

The CuriePME library is not automatically included when you install the Intel Curie board libraries. So you'll have to make sure you download it from GitHub and install it manually as follows:

1. Download the zip file from its GitHub repository (Figure 5.1): *https://github.com/01org/Intel-Pattern-Matching-Technology.*

FIGURE 5.1: Curie PME GitHub repository

2. In the Arduino IDE, choose Sketch > Include Library > Add .ZIP Library (Figure 5.2).

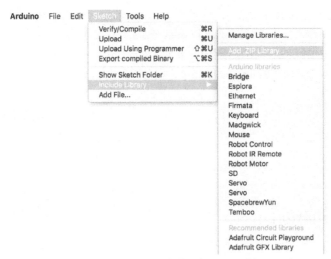

FIGURE 5.2: Selecting Sketch > Include Library > Add .ZIP Library

3. Upload the `Intel-Pattern-Matching-Technology-master.zip` file
   (Figure 5.3), and the Arduino IDE will display a "Library added
   to your libraries" message in the sketch window (Figure 5.4).

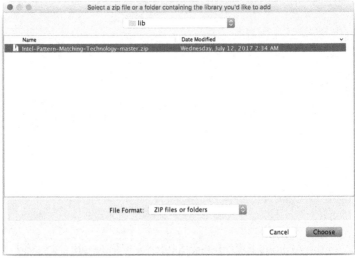

FIGURE 5.3: Choosing the `Intel-Pattern-Matching-Technology-master.zip` file

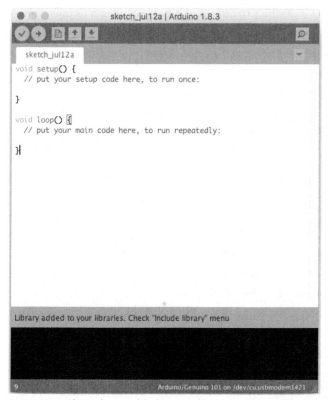

FIGURE 5.4: The "Library added to your libraries" message

4. If the previous steps are completed successfully, you should find the CuriePME library in the Arduino IDE under Sketch > Included Library > Contributed Libraries > CuriePME.

THE DRAWINGINTHEAIR EXAMPLE

The *DrawingInTheAir* example sketch (*https://github.com/01org/ Intel-Pattern-Matching-Technology/tree/master/examples/ DrawingInTheAir*) is a cool demonstration of the things possible with the CuriePME library. It allows you to draw letters in the air, using your Arduino 101 as an imaginary pen, and can recognize which letter is being drawn by analyzing patterns from the IMU data.

The URL in the previous paragraph includes a detailed tutorial from Intel in the README file. I recommend that you read it before proceeding with the experiment in this chapter, because a lot of things we will be talking about are directly based on this experiment. The tutorial will help you understand the basics of pattern recognition: *learning* and *classification*, as well as methods to improve the fidelity of the pattern matching.

PROJECT: DIY BLE GESTURE RECOGNITION MEDIA CONTROLLER

In this project, you will use the CuriePME library to build a gesture-based media controller (or a remote control). The idea will be to use this remote control to play, pause, shuffle music, change the volume, and so forth, all through gestures. For the purposes of this book, our example is limited to playing and pausing a music file.

When you first run the sketch, the Arduino 101 will ask you to make two different gestures. They could be any gestures you want, as long as they are different. For simplicity's sake, let's say that you'll draw the letters A and B in the air. You'll draw each letter several times so that the 101 can "learn" the gesture for each letter.

Then you'll run a web application in your browser that will connect via Bluetooth LE as a central device to the 101. The web app will also subscribe to a characteristic on your 101 that will hold the identified gesture. When you make the first gesture with the 101, it will recognize the motion and update its characteristic. The web app will be notified of this, and it will start playing music. When you make the second gesture with the 101, a similar process ensues and the web app pauses music that is playing. You will also be using a button that will tell the 101 to start recording gestures and an LED to make the remote offer some visual feedback to your interactions.

What You'll Need

You'll need the following (Figure 5.5):

* An Arduino 101 (with USB cable)

* A 9V battery

* A battery jack

* A whiteboard

* Some wires

* An LED

* 220 Ω resistor

* A pushbutton

* A BLE-enabled computer with the Google Chrome web browser installed on it

* Optionally, a stuffed toy, an enclosure of your choice, or some rubber bands

FIGURE 5.5: What you'll need for the "DIY BLE Gesture Recognition Media Controller" project

Circuit Assembly

Connect your LED across pin 13 and GND, as shown in Figure 5.6. Additionally, you have to connect a switch on pin 4 to start collecting IMU data when you press the switch.

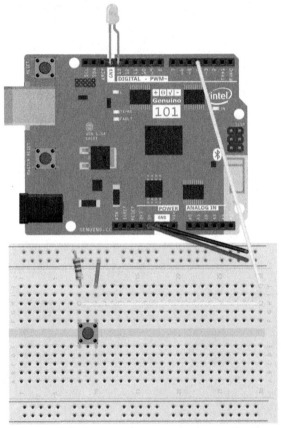

FIGURE 5.6: The circuit and component diagram for this project

System Diagram

Figure 5.7 shows the system diagram for this project.

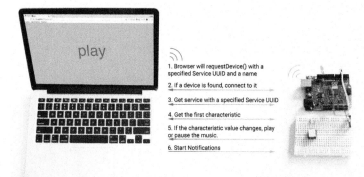

FIGURE 5.7: System diagram for our project

Arduino Sketch

You can find the complete code here: *https://github.com/ yining1023/Jumpstarting-the-Arduino-101*.

The Arduino sketch in this project is directly based on the DrawingInTheAir example. The gesture training part of your sketch will happen over the serial connection. But you want your remote to be wireless when it's being used to detect gestures, and so you will be setting up the 101 as a BLE peripheral, just as you did in the previous two chapters.

You start by defining your BLE service and initializing some global variables, as per the inline comments. vector, category, letter, and prevLetter are some global variables that will be used throughout the sketch.

```
/*
 * This example demonstrates using the pattern matching engine
(CuriePME)
 * to classify streams of accelerometer data from CurieIMU.
 *
 * It is based on the DrawingInTheAir Example by Intel Corporation.
 *
 */

#include "CurieIMU.h"
#include "CuriePME.h"
```

```
#include <CurieBLE.h>

BLEPeripheral blePeripheral;
BLEService customService("19B10000-E8F2-537E-4F6C-D104768A1216");

BLEUnsignedCharCharacteristic CharacteristicPattern(
  "4227f3b1-d6a2-4fb2-a916-3bee580a9c84",
  BLERead | BLENotify
);

/* This controls how many times a letter must be drawn during
training.
 * Any higher than 4, and you may not have enough neurons for all
26 letters
 * of the alphabet. Lower than 4 means less work for you to train
a letter,
 * but the PME may have a harder time classifying that letter. */
const unsigned int trainingReps = 4;

/* Increase this to 'A-Z' if you like-- it just takes a lot longer
to train */
const unsigned char trainingStart = 'A';
const unsigned char trainingEnd = 'B';

/* The input pin used to signal when a letter is being drawn- you'll
 * need to make sure a button is attached to this pin */
const unsigned int buttonPin = 4;
const int ledPin = 13;

/* Sample rate for accelerometer */
const unsigned int sampleRateHZ = 200;

/* No. of bytes that one neuron can hold */
const unsigned int vectorNumBytes = 128;

/* Number of processed samples (1 sample == accel x, y, z)
 * that can fit inside a neuron */
const unsigned int samplesPerVector = (vectorNumBytes / 3);

/* This value is used to convert ASCII characters A-Z
 * into decimal values 1-26, and back again. */
const unsigned int upperStart = 0x40;

const unsigned int sensorBufSize = 2048;
const int IMULow = -32768;
```

```
const int IMUHigh = 32767;

byte vector[vectorNumBytes];
unsigned int category;
char letter;
char prevLetter;
```

The setup function is where you'll begin training the 101 with the IMU data to learn two gestures. But first, as always, you want to set up and configure the BLE peripheral and set up a service and a characteristic on it. You should be familiar with this process by now.

The while loop in the setup will prevent the code after it from running until Serial is available. This way, you can be sure that Serial is initialized before you call beginTraining. The beginTraining function is the equivalent of the trainLetters function from the DrawingInTheAir example. You'll define it in a moment.

```
void setup()
{
  Serial.begin(9600);

  pinMode(buttonPin, INPUT);
  pinMode(ledPin, OUTPUT);

  // setup ble
  blePeripheral.setLocalName("CuriePME");
  blePeripheral.setAdvertisedServiceUuid(customService.uuid());
  blePeripheral.addAttribute(customService);
  blePeripheral.addAttribute(CharacteristicPattern);

  CharacteristicPattern.setValue('0');
  blePeripheral.begin();

  /* Start the IMU (Inertial Measurement Unit), enable the
accelerometer */
  CurieIMU.begin(ACCEL);

  /* Start the PME (Pattern Matching Engine) */
  CuriePME.begin();
```

```
CurieIMU.setAccelerometerRate(sampleRateHZ);
CurieIMU.setAccelerometerRange(2);

while(!Serial);
beginTraining();
Serial.println("Training complete. Now, make a gesture (remember
to ");
Serial.println("hold the button) and see if the PME can classify
it.");
Serial.println("Use another BLE device to connect to Arduino101
first");
```

```
}
```

In `loop`, you need to write logic that will enable the Curie module to guess or classify a given gesture from its knowledge and initial learning.

You want to record and guess a gesture only if the button is depressed—which is why you won't be doing anything except turning off your LED on every call to the `loop()` function when the button is not pressed.

If the button is pressed, you will read the accelerometer data from the IMU (with `readVectorFromIMU()`) and ask the CuriePME library to classify the gesture into a known category (a `letter` in this case). You also want to write that category to your characteristic pattern.

Since this process happens in a loop, you want to avoid too many unnecessary updates to `CharacteristicPattern`. So you will update it only when the value of the guessed `letter` changes from its previous value, which as you will see momentarily, is stored in the `prevLetter` variable.

```
void loop ()
{
  BLECentral central = blePeripheral.central();

  // turn off LED by default
  digitalWrite(ledPin, LOW);
```

```
// if button is pressed, read and classify the
// accelerometer vector from the IMU
if(digitalRead(buttonPin) == HIGH) {
  readVectorFromIMU(vector);
  classify(vector);
}

// if a change in the gesture or letter is
// detected, update the characteristic
if (prevLetter != letter) {
  if(central) {
    if (central.connected()) {
      CharacteristicPattern.setValue(letter);
      prevLetter = letter;
    }
  }
}
}
```

The next two functions—getAverageSample and undersample are
exactly the same as described in the DrawingInTheAir example.
The tutorial offers an in-depth explanation of these functions.
The getAverageSample function is used to smooth out and average
the noisy input from the IMU. The undersample function is used
to reduce the size of the input data to bring it down to 128 bytes
without losing any distinguishing characteristics of the data.

```
/* Simple "moving average" filter, removes low noise and other small
 * anomalies, with the effect of smoothing out the data stream. */
byte getAverageSample(byte samples[], unsigned int num, unsigned int
pos, unsigned int step)
{
  unsigned int ret;
  unsigned int size = step * 2;

  if (pos < (step * 3) || pos > (num * 3) - (step * 3)) {
    ret = samples[pos];
  } else {
    ret = 0;
    pos -= (step * 3);
    for (unsigned int i = 0; i < size; ++i) {
      ret += samples[pos - (3 * i)];
    }
```

```
    ret /= size;
  }

  return (byte)ret;
}

/* We need to compress the stream of raw accelerometer data
 * into 128 bytes, so it will fit into a neuron,
 * while preserving as much of the original pattern as possible.
 * Assuming there will typically be 1-2 seconds worth of accelerometer
 * data at 200Hz, we will need to throw away over
 * 90% of it to meet that goal!
 *
 * This is done in 2 ways:
 *
 * 1. Each sample consists of 3 signed 16-bit values
 *    (one each for X, Y and Z). Map each 16 bit value to a range of
 *    0-255 and pack it into a byte, cutting sample size in half.
 *
 * 2. Undersample. If we are sampling at 200Hz and the button is held
 *    for 1.2 seconds, then we'll have ~240 samples. Since we know
 now
 *    that each sample, once compressed, will occupy 3 of our
 neuron's
 *    128 bytes (see #1), then we know we can only fit 42 of those
 240
 *    samples into a single neuron (128 / 3 = 42.666). So if we take
 *    (for example) every 5th sample until we have 42, then we should
 *    cover most of the sample window and have some semblance of the
 *    original pattern. */
void undersample(byte samples[], int numSamples, byte vector[])
{
  unsigned int vi = 0;
  unsigned int si = 0;
  unsigned int step = numSamples / samplesPerVector;
  unsigned int remainder = numSamples - (step * samplesPerVector);

  /* Centre sample window */
  samples += (remainder / 2) * 3;
  for (unsigned int i = 0; i < samplesPerVector; ++i) {
    for (unsigned int j = 0; j < 3; ++j) {
      vector[vi + j] = getAverageSample(samples, numSamples, si + j,
step);
    }
```

```
  si += (step * 3);
  vi += 3;
  }
}
```

Now you define the `beginTraining` and `trainGesture` functions. The `beginTraining` function is called in `setup`, and it iterates
through the number of gestures you want to collect and trains
each gesture.

```
void beginTraining()
{
  for (char i = trainingStart; i <= trainingEnd; ++i) {
    Serial.print("Hold down the button and make a gesture or '");
    Serial.print(String(i) + "' in the air. Release the button as
soon ");
    Serial.println("as you are done.");

    trainGesture(i, trainingReps);
    Serial.println("OK, finished with this gesture.");
    delay(2000);
  }
}
```

The `trainGesture` function is called for each gesture to be trained.
It reads the accelerometer data (by calling `readVectorFromIMU`) from
the IMU and uses the CuriePME library to learn a gesture.

```
void trainGesture(char letter, unsigned int repeat)
{
  unsigned int i = 0;

  while (i < repeat) {
    byte vector[vectorNumBytes];

    if (i) Serial.println("And again...");

    readVectorFromIMU(vector);
    CuriePME.learn(vector, vectorNumBytes, letter - upperStart);

    Serial.println("Got it!");
    delay(1000);
```

```
    ++i;
  }
}
```

You are calling the `readVectorFromIMU` function during `setup` and `loop` to read the accelerometer data from the IMU and store it in a global variable.

In this function, when the button is pressed, the `while` loops here will keep running and prevent the `undersample` call below it from running until the button is pressed and released. This allows you to record samples from the accelerometer data continuously.

Once you finish recording a gesture, you need to `undersample` it in order to reduce the size of all the input samples you just recorded. When you have an undersampled input, you can run it through either a learn or a classify function to either train or guess the recorded gesture.

```
void readVectorFromIMU(byte vector[])
{
  byte accel[sensorBufSize];
  int raw[3];

  unsigned int samples = 0;
  unsigned int i = 0;

  /* Wait until button is pressed */
  while (digitalRead(buttonPin) == LOW) {
    digitalWrite(ledPin, LOW);
  }

  /* While button is being held... */
  while (digitalRead(buttonPin) == HIGH) {
    digitalWrite(ledPin, HIGH);
    if (CurieIMU.dataReady()) {
      CurieIMU.readAccelerometer(raw[0], raw[1], raw[2]);

      /* Map raw values to 0-255 */
      accel[i] = (byte) map(raw[0], IMULow, IMUHigh, 0, 255);
      accel[i + 1] = (byte) map(raw[1], IMULow, IMUHigh, 0, 255);
      accel[i + 2] = (byte) map(raw[2], IMULow, IMUHigh, 0, 255);
```

```
    i += 3;
    ++samples;

    /* If there's not enough room left in the buffers
     * for the next read, then we're done */
    if (i + 3 > sensorBufSize) {
      break;
    }
  }
}
undersample(accel, samples, vector);
}
```

Lastly, our classify function takes an input vector and uses the CuriePME library to classify it into a known category. You are mapping your categories to letters, just like the DrawingInTheAir example.

Once CuriePME is done with the classification process, it updates the global variable letter with the newly classified gesture. Note that before updating letter you copy its previous value to the variable prevLetter, which is used in the loop function.

```
void classify(byte vector[])
{
  /* Use the PME to classify the vector, i.e. return a category from
1-26,
   * representing a letter from A-Z */
  category = CuriePME.classify(vector, vectorNumBytes);
  Serial.println("get category: ");
  Serial.println(category);
  prevLetter = letter;
  if (category == CuriePME.noMatch) {
    Serial.println("Don't recognise that one-- try again.");
    letter = '0';
  } else {
    letter = category + upperStart;
    Serial.println("The gesture is:");
    Serial.println(letter);
  }
}
```

TESTING THE ARDUINO CODE

Once you are done setting up the circuit and uploading the sketch to your 101, open the serial monitor while keeping your Arduino connected to your computer via the USB cable. You can see that the program prompts you to press the button and record a gesture multiple times. As stated in our code, it should ask you to record two gestures (multiple tries per gesture).

> **NOTE** When you are training the Arduino 101, keep it connected to both your computer and a 9V battery. The battery will provide power to the 101 after you disconnect the USB cable in the next section—you don't want the sketch to restart after you disconnect the USB cable. If you want to persist the training data into a knowledge base, take a look at some of the other examples in the CuriePME repository: *https://github.com/01org/Intel-Pattern-Matching-Technology#restoring-knowledge*.

FIGURE 5.8: Setup for training through the serial port

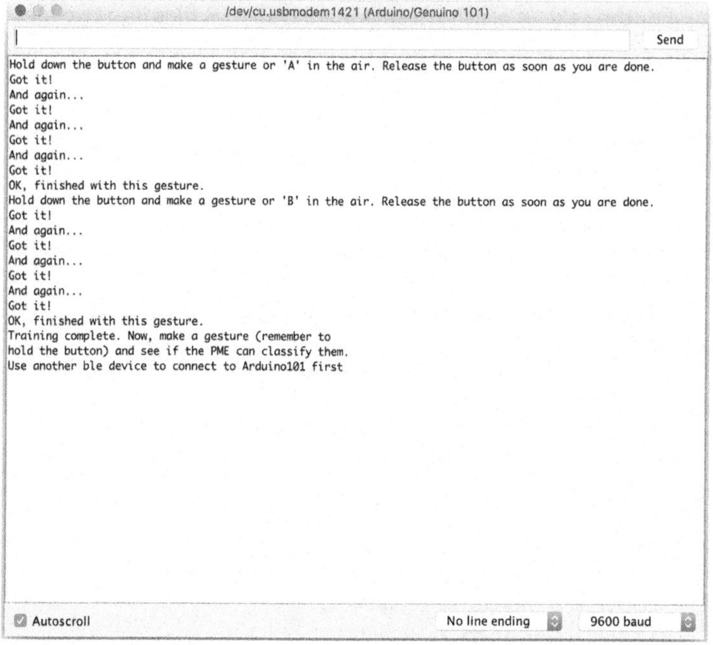
/dev/cu.usbmodem1421 (Arduino/Genuino 101)
Send
Hold down the button and make a gesture or 'A' in the air. Release the button as soon as you are done.
Got it!
And again...
Got it!
And again...
Got it!
And again...
Got it!
OK, finished with this gesture.
Hold down the button and make a gesture or 'B' in the air. Release the button as soon as you are done.
Got it!
And again...
Got it!
And again...
Got it!
And again...
Got it!
OK, finished with this gesture.
Training complete. Now, make a gesture (remember to
hold the button) and see if the PME can classify them.
Use another ble device to connect to Arduino101 first
Autoscroll
No line ending
9600 baud

FIGURE 5.9: The Arduino IDE serial port results after training

Once you are done recording gestures, run your web app. But first let's walk through the JavaScript, HTML, and CSS for the web app.

Web App Code

Our web app this time will consist of four files: `index.html`, `main.js`, `ble.js`, and `styles.css`.

index.html

The HTML for this app will largely be the same as the previous chapter. The only difference is that I have removed a separate "Start Notifications" button. This code will start notifications by default after you connect to the peripheral.

For the sake of clarity, I'll split our JavaScript files into two files, `main.js` and `ble.js`, and import them separately in this HTML file.

```
<!doctype html>
<html lang="en">
  <head>
    <meta charset="utf-8">
    <title>BLE</title>
    <link href="https://fonts.googleapis.com/css?family=Roboto"
rel="stylesheet">
    <script src="https://cdnjs.cloudflare.com/ajax/libs/p5.js/0.5.4/
p5.min.js"></script>
    <script src="https://cdnjs.cloudflare.com/ajax/libs/p5.js/0.5.4/
addons/p5.sound.min.js"></script>
    <link rel="stylesheet" type="text/css" href="style.css">
  </head>

  <body>
    <button onClick="connect()">Connect</button>
    <button onClick="disconnect()">Disconnect</button>
    <script src="ble.js"></script>
    <script src="main.js"></script>
  </body>
</html>
```

ble.js

As before, you'll store your service UUID and name in variables. You'll also define some variables that you will be using throughout.

```
// our service UUID and name
const serviceUuid = '19b10000-e8f2-537e-4f6c-d104768a1216';
const name = 'CuriePME';

var bluetoothDevice, characteristicPattern, value;
var pattern = '';
```

The connect function is also exactly the same as in Chapter 4, but with one small change. This time, you also use `startNotifications` after you get the characteristics to avoid having to click another button to start it.

```
function connect() {
  let options = {
    filters: [{
      services: [serviceUuid],
      name: name
    }]}

  console.log('Requesting Bluetooth Device...');

  navigator.bluetooth.requestDevice(options)
    .then(device => {
      bluetoothDevice = device; // save a copy
      console.log('Got device', device.name);
      return device.gatt.connect();
    })
    .then(server => {
      console.log('Getting Service...');
      return server.getPrimaryService(serviceUuid);
    })
    .then(service => {
      console.log('Getting Characteristics...');
      // Get all characteristics.
      return service.getCharacteristics();
    })
    .then(characteristics => {
      console.log('Got Characteristics');
      console.log('Starting Notifications...');

      // select the first characteristic
      characteristicPattern = characteristics[0];

      // add event listener to handle characteristic value change
      characteristicPattern.addEventListener(
        'characteristicvaluechanged',
        handleData
      );
      // start notifications
      return characteristicPattern.startNotifications();
    })
    .catch(error => {
      console.log('Argh! ' + error);
    });
}
```

The `handleData` function will get the received data from `event.target`. This data is a buffer value, so you have to convert it to an integer by calling the `getUint8` function. You further convert the integer `value` to a character. Although this step is not necessary, it makes the code easier to read because you can compare two basic characters (in this case, A and B) rather than their ASCII values.

```
function handleData(event) {
  value = event.target.value.getUint8(0);
  console.log('> Got Pattern data: ' + value);
  if (value !== 48) {
    pattern = String.fromCharCode(value);
    console.log('Received Pattern ' + pattern);
  } else {
    pattern = null;
    console.log('Could not recognize the gesture...');
  }
}
```

Finally, you also write a `disconnect` function, which will be called when you click the Disconnect button. This function will close the Bluetooth connection.

```
function disconnect() {
  if (bluetoothDevice && bluetoothDevice.gatt) {
    bluetoothDevice.gatt.disconnect();
    console.log('Disconnected');
  }
}
```

main.js

Most of the p5.js code related to updating the rendered view will live in the `main.js` file. Here you define variables and background colors that will change depending on the gesture received.

```
var song;
var displayText = 'Waiting...';
var backgroundColors = ['#1DFFAD', '#0FE8E2'];
```

Start by asynchronously loading a song file of your choice in the preload function. For this example, we have saved our song.mov file in the same folder as main.js. In the setup function, you'll create a canvas and set up some basic styles.

```
function preload() {
  song = loadSound('./song.mov');
}
function setup() {
  createCanvas(windowWidth, windowHeight);
  textSize(180);
  textAlign(CENTER);
  noStroke();
  fill('#FF5797');
  song.setVolume(0.5);
}
```

In the draw loop, you'll call the function patternMatched if a valid pattern is recognized. This function will be responsible for playing or pausing the song, changing the background color, and displaying the text.

```
function draw() {
  if (pattern !== '') {
    patternMatched();
  }
  text(displayText, width / 2, height / 2);
}
```

Finally, in the patternMatched function, you check whether the pattern found is A or B. For A the song plays; for B, it pauses. Depending on the recognized pattern, the patternMatched function also changes the background color to one of the colors from the backgroundColors array.

```
function patternMatched() {
  // if pattern A is recognized, play the song
  // else pause it
  if (pattern === 'A') {
    displayText = 'play';
    background(backgroundColors[0]);
    if (!song.isPlaying()) {
```

```
      console.log('Playing Song...');
      song.play();
    }
  } else if (pattern === 'B') {
    displayText = 'pause';
    background(backgroundColors[1]);
    if (song.isPlaying()) {
      console.log('Pausing Song...');
      song.pause();
    }
  }
}
```

style.css

Finally, you want some basic styles to be used in your app.

```css
html, body {
  overflow: hidden;
  margin: 0;
  padding: 0;
  font-family: 'Roboto', sans-serif;
}
```

RUNNING THE APP

Make sure you trained your Arduino with a USB cable and a battery connected to it, as explained in the previous section. Once you are done, you can disconnect the Arduino from the USB cable and the battery will provide power to the 101. This will allow you to go wireless!

Then, in a Terminal window, cd into your app directory and start your Python server. You saw a similar process in the "Running the App" section in Chapter 4.

```
cd /path/to/your/web/app/folder
```

For Python 2:

```
python -m SimpleHTTPServer
```

For Python 3:

```
python3 -m http.server
```

Then you can navigate to *localhost:8000* in your browser to see the app running. Finally, you will connect and subscribe to the characteristics on your Arduino using the buttons on the web page. This will cause the web page to start listening for changes on the gesture characteristic. If you press the button on your circuit and move your 101 in a pattern, a gesture will be recorded. Once the 101 classifies a valid gesture from the 101, the song will either play or pause, depending on the gesture you made.

FIGURE 5.10: Final setup

FIGURE 5.11: The browser: displaying the "Waiting" text while idle

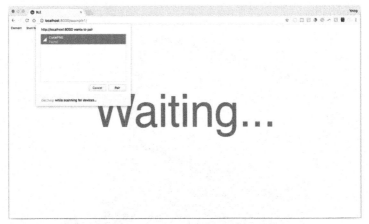

FIGURE 5.12: The browser: connecting to the Arduino 101

FIGURE 5.13: The browser: recognizing the first gesture

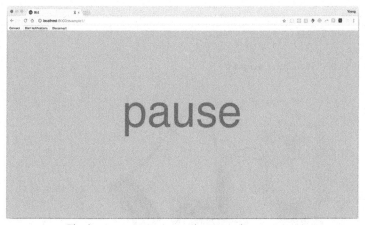

FIGURE 5.14: The browser: recognizing the second gesture

Fabrication (Optional)

As in Chapter 4, I chose to use my soft toy to enclose the circuit and put it in a usable form. I encourage you to come up with your own tactile enclosure for your remote.

FIGURE 5.15: Circuit diagram

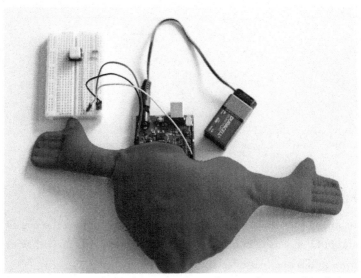

FIGURE 5.16: Connecting a battery to the Arduino and putting the whole assembly into a soft toy

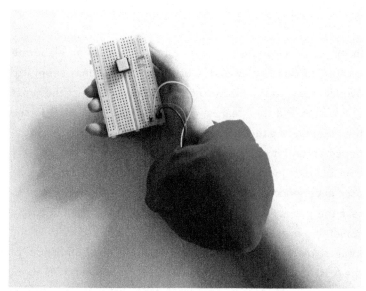

FIGURE 5.17: Wearing the soft toy on my wrist with the button accessible

FINAL RESULT

You'll find a video demonstration for training the 101 here:

https://youtu.be/lBIWzQxxPZO

You can also view a video demo of the web app here:

https://youtu.be/CuWK2fc2dCw

TROUBLESHOOTING

When you are training your 101, be consistent with your gestures during multiple trials. Remember to hold the button tightly—don't let it go until you finish your gesture. I recommend that you use a big pushbutton or an Arduino shield, because it will be much more ergonomic and help you obtain consistent results.

FURTHER EXPLORATION

In this project you had to press a button and do an elaborate gesture to simply play and pause a song. That may seem like overkill when you realize you could do that with just the press of a button. But the idea, as with every maker project, is to see the potential of a technology and design an excellent user experience around it. A gesture-based remote control sounds very cool, but designing the gestures, the form factor, and the overall user experience is a big challenge in itself. I encourage you to take this project to the next step: break it, rebuild it, and create something even better out of it. Tom Igoe, arts professor at the Interactive Telecommunications Program (ITP), Tisch School of the Arts, New York University, makes some excellent points about user experience design in his book *Making Things Talk* (Maker Media, 2017). I highly recommend that you read this book and take your connected projects a step further!

The Arduino 101 has great hardware, but what makes it really special are the different Curie libraries that have been written for it. You used some of these libraries, along with the Web Bluetooth API, to see all the different things possible with such a device. I encourage you to check out the other libraries that we did not cover in this book, such as the CurieSerialFlash library, which allows you to store data and organize persistent data on the Arduino, or the CurieTime library, which lets you do date and time manipulations on the Arduino 101. You can find some examples of projects using libraries here: *https://github.com/01org/corelibs-arduino101*. When you see what capabilities these and other libraries provide to you, you can't help but get inspired to make the most out of your projects!

I hope you enjoyed this book. If you have any questions about the book or the projects in the book, feel free to reach out to me on the GitHub repository for this book by creating an issue.

INDEX

CPSIA information can be obtained
at www.ICGtesting.com
Printed in the USA
BVOW07s0915090318
510143BV00005B/89/P